GEOMETRIC AND ENGINEERING DRAWING
for C.S.E. and G.C.E.

K. MORLING, Graduate of the Institution of
Mechanical Engineers

SI Units

Edward Arnold

© K. Morling 1974

First published 1969
by Edward Arnold (Publishers) Ltd
41 Bedford Square, London WC1B 3DQ

Edward Arnold (Australia) Pty Ltd
80 Waverley Road, Caulfield East, Victoria 3145, Australia

Reprinted 1969, 1970, 1971, 1972, 1973

Second edition 1974

Reprinted 1975, 1976, (with revisions) 1977, 1978, 1980, 1981, 1983, 1985

ISBN: 0 7131 3319 8

For Jilly

Printed in Great Britain by Fletcher & Son Ltd, Norwich

Preface to the second edition

Whilst there are several good introductory books about Technical Drawing, I think that there is still a need for a book written for students studying in their final years for the Secondary Certificate of Education and for the General Certificate of Education. Such a book should, within the limits of the syllabus, cover the whole range of the subject, have large diagrams free of irrelevant information and have a large number of examples of the requisite standard. Now that the book is completed, I appreciate how demanding these conditions are; I leave it to the reader to decide whether or not they have been fulfilled.

The units used in the second edition are entirely metric and the dimensioning and drawing specifications are to BS 308:1972. The change to the use of metric units has not been paralleled with a change to a standard projection. Third Angle projection is widely used in industry and this practice is spreading to schools and Examining Boards, and students are usually offered an alternative between drawing in First or Third Angle projection. Both systems are used in this book with Third Angle predominant.

I have tried to arrange the contents of this book so that each chapter is self-contained. This is alien to the nature of the discipline but for obvious reasons individual techniques have to be dealt with individually. I do not suggest that chapters should necessarily be studied consecutively but, if this course is adopted, students will find, with very few exceptions, that each chapter, whilst often assuming knowledge from previous text, will not assume knowledge from following text.

Wherever possible, the examples in this book have been taken from recent examination papers. Many of these examples were originally set in Imperial units and, where dimensions have been changed, the pattern of dimensioning is in line with recommendations for dimensioning in metric units.

However carefully a manuscript is written and checked, errors still go unnoticed. I apologize in advance for any mistakes and I shall be grateful if they are pointed out to me via the publishers.

K.M.
January, 1974

Acknowledgments

I wish to express my grateful thanks to the following examining Boards for giving their permission to use questions from past papers. I am particularly grateful for their allowing me to change many of the questions from Imperial to Metric units.

Certificate of Secondary Education
Associated Lancashire Schools Examining Board
East Anglian Regional Examinations Board
Metropolitan Regional Examinations Board
Middlesex Regional Examining Board
North Western Secondary School Examinations Board
South-East Regional Examinations Board
Southern Regional Examinations Board
West Midlands Examinations Board
General Certificate of Education
Associated Examining Board
Local Examinations Syndicate, University of Cambridge

Joint Matriculation Board
University of London
Oxford Delegacy of Local Examinations
Oxford and Cambridge Schools Examination Board
Southern Universities' Joint Board

My special thanks to the West Midlands Examinations Board, the Associated Lancashire Schools Examining Board and the Southern Universities Joint Board for allowing me to draw solutions to questions set by them (Fig. 18/37, 18/39 and 18/41).

The extracts from BS 308:1972 'Engineering Drawing Practice' and BS 3692:1967 'ISO Metric Precision Hexagon Bolts, Screws and Nuts' are reproduced by permission of the British Standards Institution, 2 Park Street, London, W.1, from whom copies of the complete standards may be obtained.

My thanks to S. Pagett for checking the manuscript.

K.M.

Contents

Part I
Geometric drawing

1

Scales

Before you start any drawing you first decide how large the drawings have to be. The different views of the object to be drawn must not be bunched together or be too far apart. If you are able to do this and still draw the object in its natural size then obviously this is best. This is not always possible; the object may be much too large for the paper or much too small to be drawn clearly. In either case it will be necessary to draw the object 'to scale'. The scale must depend on the size of the object; a miniature electronic component may have to be drawn 100 times larger than it really is, whilst some maps have natural dimensions divided by millions.

There are drawing aids called 'scales' which are designed to help the draughtsman cope with these scaled dimensions. They look like an ordinary ruler but closer inspection shows that the divisions on these scales are not the usual centimetres or millimetres, but can represent them. These scales are very useful but there will come a time when you will want to draw to a size that is not on one of these scales. You could work out the scaled size for every dimension on the drawing but this can be a long and tedious business—unless you construct your own scale. This chapter shows you how to construct any scale that you wish.

The Representative Fraction (R.F.)

The representative fraction shows instantly the ratio of the size of the line on your drawing and the natural size. The ratio of numerator to denominator of the fraction is the ratio of drawn size to natural size. Thus, a representative fraction of $\frac{1}{5}$ means that the actual size of the object is five times the size of the drawing of that object.

If a scale is given as 1 mm = 1 m then the R.F. is

$$\frac{1 \text{ mm}}{1 \text{ m}} = \frac{1 \text{ mm}}{1000 \text{ mm}} = \frac{1}{1000}.$$

A cartographer (a map draughtsman) has to work with some very large scales. He may have to find, for instance, the R.F. for a scale of 1 mm = 5 km. In this case the R.F. will be $\frac{1 \text{ mm}}{5 \text{ km}} = \frac{1}{5 \times 1000 \times 1000} = \frac{1}{5\,000\,000}.$

Plain Scales

There are two types of scales, plain and diagonal. The plain scale is used for simple scales, scales that do not have many sub-divisions.

When constructing any scale, the first thing to decide is the length of the scale. The obvious length is a little longer than the longest dimension on the drawing. Fig. 1/1 shows a very simple scale of 20 mm = 100 mm. The

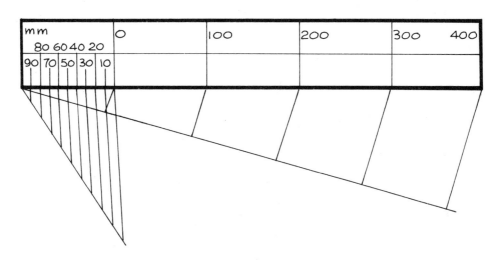

Fig. 1/1   Plain scale 10 mm = 100 mm or 1 mm = 5 mm

largest natural dimension is 500 mm so the total length of the scale is $\frac{500}{5}$ mm or 100 mm. This 100 mm is divided into 5 equal portions, each portion representing 100 mm. The first 100 mm is then divided into 10 equal portions, each portion representing 10 mm. These divisions are then clearly marked to show what each portion represents.

Finish is very important when drawing scales. You would not wish to use a badly graduated or poorly marked ruler and you should apply the same standards to your scales. Make sure that they are marked with all the important measurements.

Fig. 1/2 shows another plain scale. This one would be used where the drawn size would be three times bigger than the natural size.

To construct a plain scale, 30 mm = 10 mm, 50 mm long to read to 1 mm (Fig. 1/2)

Length of scale $= 30 \times 5 = 150$ mm
1st division $\qquad 5 \times 10$ mm
2nd division $\qquad 10 \times 1$ mm

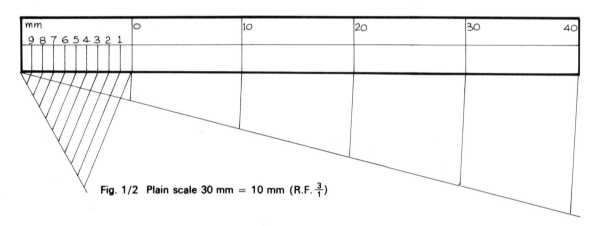

Fig. 1/2 Plain scale 30 mm = 10 mm (R.F. $\frac{3}{1}$)

Diagonal Scales

There is a limit to the number of divisions that can be constructed on a plain scale. Try to divide 10 mm into 50 parts; you will find that it is almost impossible. The architect, cartographer and surveyor all have the problem of having to sub-divide into smaller units than a plain scale allows. A diagonal scale allows you to divide into smaller units.

Before looking at any particular diagonal scale, let us first look at the underlying principle.

Fig. 1/3 shows a triangle ABC. Suppose that AB is 10 mm long and BC is divided into 10 equal parts. Lines from these equal parts have been drawn parallel to AB and numbered from 1 to 10.

It should be obvious that the line 5–5 is half the length of AB. Similarly, the line 1–1 is $\frac{1}{10}$ the length of AB and line 7–7 is $\frac{7}{10}$ the length of AB. (If you wish to prove this mathematically use similar triangles.)

You can see that the lengths of the lines 1–1 to 10–10 increase by 1 mm each time you go up a line. If the length of AB had been 1 mm to begin with the increases would have been $\frac{1}{10}$ mm each time. In this way small lengths can be divided into very much smaller lengths, and can be easily picked out.

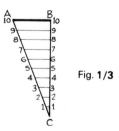

Fig. 1/3

Three examples of diagonal scales follow.

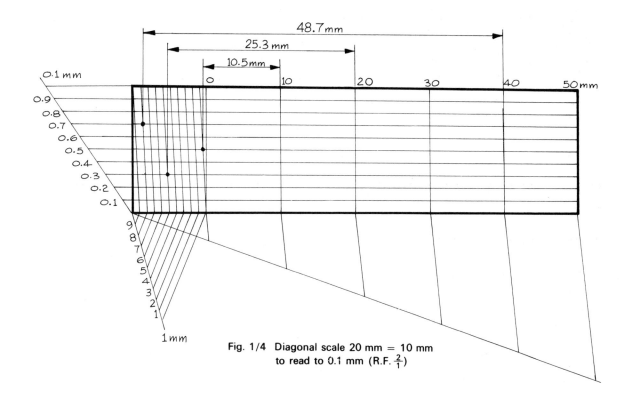

Fig. 1/4 Diagonal scale 20 mm = 10 mm
to read to 0.1 mm (R.F. $\frac{2}{1}$)

This scale would be used where the drawing is twice the size of the natural object and the draughtsman has to be able to measure on a scale accurate to 0.1 mm.

The longest natural dimension is 60 mm. This length is first divided into six 10 mm intervals. The first 10 mm is then divided into 10 parts, each 1 mm wide (scaled). Each of these 1 mm intervals is divided with a diagonal into 10 more equal parts.

5

To construct a diagonal scale, 30 mm = 1 mm,
4 m long to read to 10 mm Fig 1/5

Length of scale is 4 × 30 mm = 120 mm.
First division into 4   30 mm lengths.
Second division into 10   100 mm lengths.
Diagonal division into 10   10 mm lengths.

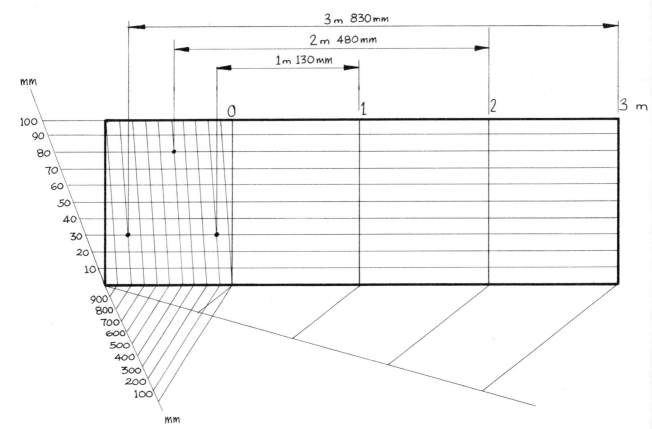

Fig. 1/5 A diagonal scale, 3 cm = 1 m, to read m and cm

To construct a diagonal scale, 50 mm = 1 mm, 3 mm long to read 0.01 mm Fig. 1/6

Length of scale is 3×50 mm = 150 mm.
First division is 3   50 mm lengths.
Second division is 10   0.1 mm lengths.
Diagonal division is 10   0.01 mm lengths.

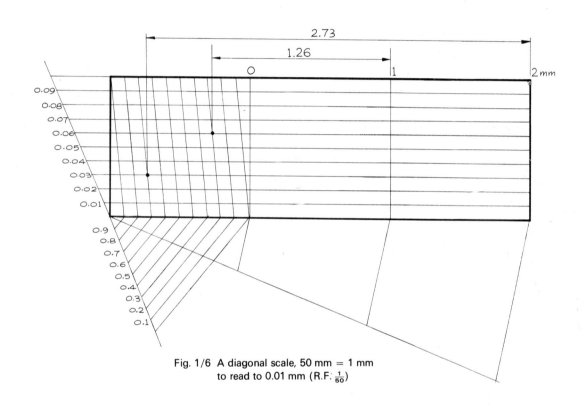

Fig. 1/6 A diagonal scale, 50 mm = 1 mm
to read to 0.01 mm (R.F. $\frac{1}{50}$)

Proportional Scales

It is possible to construct one plain scale directly from another, so that the new scale is proportional to the original one. An example of this is given in Fig. 1/7. The new scale is a copy of the original one but is $\frac{7}{4}$ times larger. The proportions of the scales can be varied by changing the ratios of the lines AB to BC.

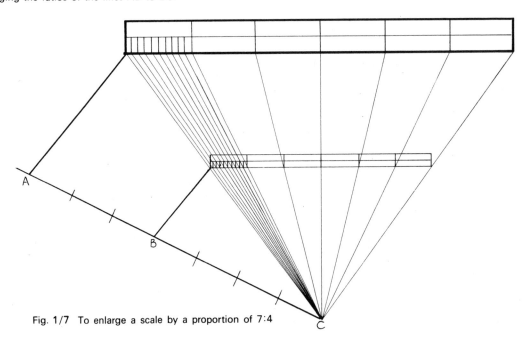

Fig. 1/7 To enlarge a scale by a proportion of 7:4

Exercises 1

1. (a) Draw the simple key shown in Fig. 1 *full size.*

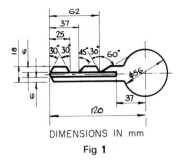

DIMENSIONS IN mm

Fig 1

(b) Construct a plain scale with a representative fraction of $\frac{5}{4}$, suitable for use in the making of an enlarged drawing of this key. Do *not* draw this key again.
Southern Regional Examinations Board (Question originally set in Imperial units).

2. Construct a plain scale of 50 mm = 300 mm to read to 10 mm up to 1200 mm. Using this scale, draw to scale a triangle having a perimeter of 1200 mm and having sides in the ratio 3:4:6. Print neatly along each side the length to the nearest 10 mm.
Oxford Local Examinations (Question originally set in Imperial units; see Chapter 2 for information not in Chapter 1).

3. Construct the plain figure shown in Fig. 2 and then, by means of a proportional scale, draw a similar figure standing on the base AG. All angles must be constructed geometrically in the first figure. Measure and state the length of the side corresponding with CD.
Oxford Local Examinations (Question originally set in Imperial units; see Chapter 2 for information not in Chapter 1).

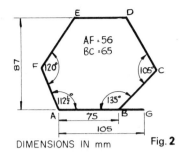

AF = 56
BC = 65

87

DIMENSIONS IN mm   Fig. **2**

4. Construct a diagonal scale in which 40 mm represents 1 m. The scale is to read down to 10 mm and is to cover a range of 5 m. Mark off a distance of 4 m 780 mm.

5. Construct a diagonal scale of 25 mm to represent 1 m which can be used to measure m and 10 mm up to 8 m. Using this scale construct a quadrilateral ABCD which stands on a base AB of length 4 m 720 mm and having BC = 3 m 530 mm, AD = 4 m 170 mm, $\angle$ABC = 120° and $\angle$ADC = 90°. Measure and state the lengths of the two diagonals and the perpendicular height, all correct to the nearest 10 mm. Angles must be constructed geometrically.

   Oxford Local Examinations (Question originally set in Imperial units; see Chapter 2 for information not in Chapter 1).

6. Construct a diagonal scale, ten times full size, to show mm and tenths of a mm and to read to a maximum of 20 mm. Using the scale, construct a triangle ABC with AB 17.4 mm, BC 13.8 mm and AC 11 mm.
   Oxford and Cambridge Schools Examinations Board.

2

The construction of geometric figures from given data

This chapter is concerned with the construction of plane geometric figures. Plane geometry is the geometry of figures that are two-dimensional, i.e. figures that have only length and breadth. Solid geometry is the geometry of three-dimensional figures.

There are an endless number of plane figures but we will concern ourselves only with the more common ones —the triangle, the quadrilateral and the better known polygons.

Before we look at any particular figure, there are a few constructions that must be revised.

Fig. 2/1 To construct a parallel line

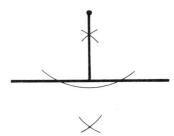

Fig. 2/4 To erect a perpendicular from a point to a line

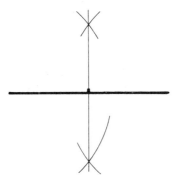

Fig. 2/2 To bisect a line

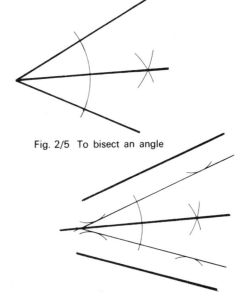

Fig. 2/5 To bisect an angle

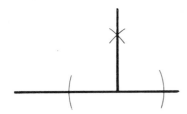

Fig. 2/3 To erect a perpendicular from a point on a line

Fig. 2/6 To bisect the angle formed by two converging lines

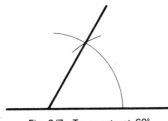

Fig. 2/7 To construct 60°

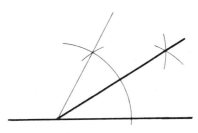

Fig. 2/8 To construct 30°

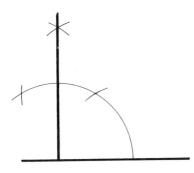

Fig. 2/9 To construct 90°

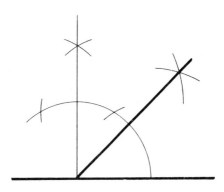

Fig. 2/10 To construct 45°

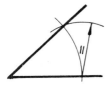 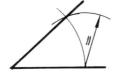

Fig. 2/11 To construct an angle
similar to another angle

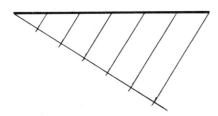

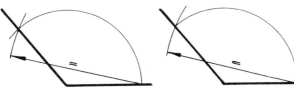

Fig. 2/12 To divide a line into a number
of equal parts (say 6)

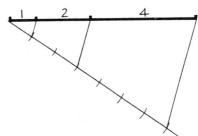

Fig. 2/13 To divide a line
proportionally (say 1:2:4)

11

THE TRIANGLE

Definitions

The triangle is a plane figure bounded by three straight sides.

A *scalene* triangle is a triangle with three unequal sides and three unequal angles.

An *isosceles* triangle is a triangle with two sides, and hence two angles, equal.

An *equilateral* triangle is a triangle with all the sides, and hence all the angles, equal.

A *right-angled* triangle is a triangle containing one right angle. The side opposite the right-angle is called the hypotenuse.

Constructions

To construct an equilateral triangle, given one of the sides (Fig. 2/14)

1. Draw a line AB, equal to the length of the side.
2. With compass point on A and radius AB, draw an arc as shown.

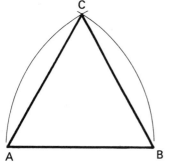

Fig. 2/14

3. With compass point on B, and with the same radius, draw another arc to cut the first arc at C.

Triangle ABC is equilateral.

To construct an isosceles triangle given the perimeter and the altitude (Fig. 2/15)

1. Draw line AB equal to half the perimeter.
2. From B erect a perpendicular and make BC equal to the altitude.
3. Join AC and bisect it to cut AB in D.

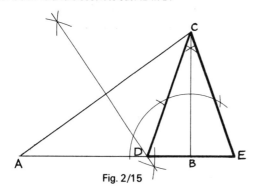

Fig. 2/15

4. Produce DB so that BE = BD.

CDE is the required triangle.

To construct a triangle, given the base angles and the altitude (Fig. 2/16)

1. Draw a line AB.
2. Construct CD parallel to AB so that the distance between them is equal to the altitude.
3. From any point E, on CD, draw CÊF and DÊG so that they cut AB in F and G respectively.

Since CÊF = EF̂G and DÊG = EĜF (alternate angles), then EFG is the required triangle.

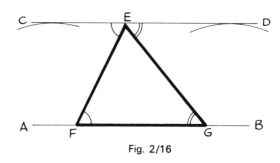

Fig. 2/16

To construct a triangle given the base, the altitude and the vertical angle (Fig. 2/17)

1. Draw the base AB.
2. Construct BÂC equal to the vertical angle.
3. Erect AD perpendicular to AC.
4. Bisect AB to meet AD in 0.
5. With centre O and radius OA (= OB), draw a circle.
6. Construct EF parallel to AB so that the distance between them is equal to the altitude.

Let EF intersect the circle in G.

ABG is the required triangle.

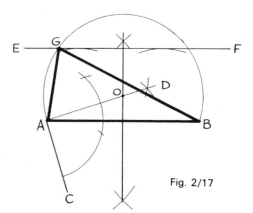

Fig. 2/17

To construct a triangle given the perimeter and the ratio of the sides (Fig. 2/18)

1. Draw the line AB equal in length to the perimeter.
2. Divide AB into the required ratio (say 4 : 3 : 6).
3. With centre C and radius CA draw an arc.
4. With centre D and radius DB draw an arc to intersect the first arc in E.

ECD is the required triangle.

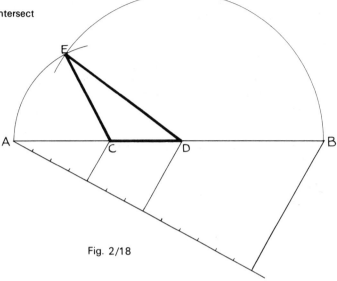

Fig. 2/18

To construct a triangle given the perimeter, the altitude and the vertical angle (Fig. 2/19)

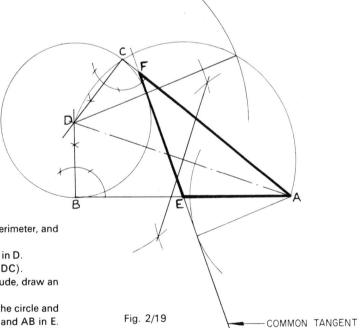

1. Draw AB and AC each equal to half the perimeter, and so that CAB is the vertical angle.
2. From B and C erect perpendiculars to meet in D.
3. With centre D, draw a circle, radius DB (= DC).
4. With centre A and radius equal to the altitude, draw an arc.
5. Construct the common tangent between the circle and the arc. Let this tangent intersect AC in F and AB in E. (For tangent construction see Chapter 4.)

FEA is the required triangle.

Fig. 2/19

COMMON TANGENT

13

To construct a triangle similar to another triangle but with a different perimeter (Fig. 2/20)

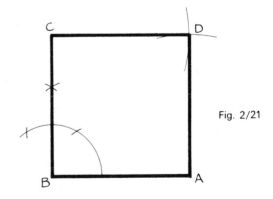

Fig. 2/20

1. Draw the given triangle ABC.
2. Produce BC in both directions.
3. With compass point on B and radius BA, draw an arc to cut CB produced in F.
4. With compass point on C and radius CA, draw an arc to cut BC produced in E.
5. Draw a line FG equal in length to the required perimeter.
6. Join EG and draw CJ and BH parallel to it.
7. With centre H and radius HF draw an arc.
8. With centre J and radius JG draw another arc to intersect the first arc in K.

HKJ is the required triangle.

THE QUADRILATERAL

Definitions

The quadrilateral is a plane figure bounded by four straight sides.

A *square* is a quadrilateral with all four sides of equal length and one of its angles (and hence the other three) a right angle.

A *rectangle* is a quadrilateral with its opposite sides of equal length and one of its angles (and hence the other three) a right angle.

A *parallelogram* is a quadrilateral with opposite sides equal and therefore parallel.

A *rhombus* is a quadrilateral with all four sides equal.

A *trapezium* is a quadrilateral with one pair of opposite sides parallel.

A *trapezoid* is a quadrilateral with all four sides and angles unequal.

Constructions

To construct a square given the length of the side (Fig. 2/21)

1. Draw the side AB.
2. From B erect a perpendicular.
3. Mark off the length of side BC.
4. With centres A and C draw arcs, radius equal to the length of the side of the square, to intersect at D.

ABCD is the required square.

Fig. 2/21

To construct a square given the diagonal (Fig. 2/22)
1. Draw the diagonal AC.
2. Bisect AC.
3. With centre O and radius OA (= OC) draw a circle to cut the bisecting line in B and D.

ABCD is the required square.

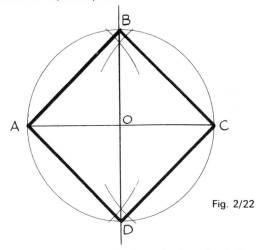

Fig. 2/22

To construct a rectangle given the length of the diagonal and one of the sides (Fig. 2/23)
1. Draw the diagonal BD.
2. Bisect BD.
3. With centre O and radius OB (= OD) draw a circle.
4. With centre B and radius equal to the length of the known side, draw an arc to cut the circle in C.
5. Repeat step 4 with centre D to cut at A.

ABCD is the required rectangle.

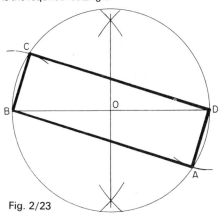

Fig. 2/23

To construct a parallelogram given two sides and an angle (Fig. 2/24)
1. Draw AD equal to the length of one of the sides.
2. From A construct the known angle.
3. Mark off AB equal in length to the other known side.
4. With compass point at B draw an arc equal in radius to AD.

5. With compass point at D draw an arc equal in radius to AB.

ABCD is the required parallelogram.

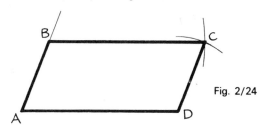

Fig. 2/24

To construct a rhombus given the diagonal and the length of the sides (Fig. 2/25)
1. Draw the diagonal AC.
2. From A and C draw intersecting arcs, equal in length to the sides, to meet at B and D.

ABCD is the required rhombus.

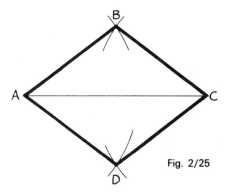

Fig. 2/25

To construct a trapezium given the lengths of the parallel sides, the perpendicular distance between them and one angle (Fig. 2/26)
1. Draw one of the parallels AB.
2. Construct the parallel line.
3. Construct the known angle from B to intersect the parallel line in C.
4. Mark off the known length CD.

ABCD is the required trapezium.

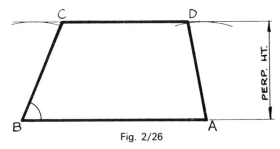

Fig. 2/26

15

POLYGONS

Definitions

A polygon is a plane figure bounded by more than four straight sides. Polygons that are frequently referred to have particular names. Some of these are listed below.

A *pentagon* is a plane figure bounded by five sides.

A *hexagon* is a plane figure bounded by six sides.

A *heptagon* is a plane figure bounded by seven sides.

An *octagon* is a plane figure bounded by eight sides.

A *nonagon* is a plane figure bounded by nine sides.

A *decagon* is a plane figure bounded by ten sides.

A regular polygon is one that has all its sides equal and therefore all its exterior angles equal and all its interior angles equal.

It is possible to construct a circle within a regular polygon so that all the sides of the polygon are tangential to that circle. The diameter of that circle is called the diameter of the polygon. If the polygon has an even number of sides, the diameter is the distance between two diametrically opposed faces. This dimension is often called the 'across-flats' dimension.

The diagonal of a polygon is the distance from one corner to the corner furthest away from it. If the polygon has an even number of sides, then this distance is the dimension between two diametrically opposed corners.

Constructions

To construct a regular hexagon given the length of the sides (Fig. 2/27)

1. Draw a circle, radius equal to the length of the side.
2. From any point on the circumference, step the radius around the circle six times. If your construction is accurate, you will finish at exactly the same place that you started.
3. Connect the six points to form a regular hexagon.

To construct a regular hexagon given the diameter (Fig. 2/28)

This construction, using compasses and straight edge only, is quite feasible but is relatively unimportant. What is important is to recognize that a hexagon can be constructed, given the diameter or across-flats dimension, by drawing tangents to the circle with a 60° set square. This is very important when drawing hexagonal-headed nuts and bolts.

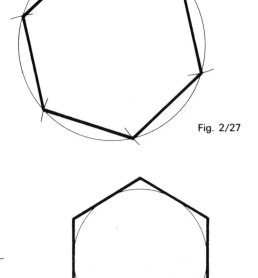

Fig. 2/27

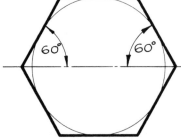

Fig. 2/28

To construct a regular octagon given the diagonal, i.e. within a given circle (Fig. 2/29)

1. Draw the circle and insert a diameter AE.
2. Construct another diagonal CG, perpendicular to the first diagonal.
3. Bisect the four quadrants thus produced to cut the circle in B, D, F, and H.

ABCDEFGH is the required octagon.

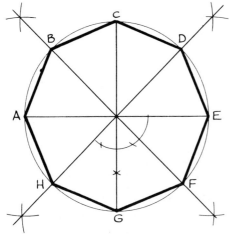

Fig. 2/29

To construct a regular octagon given the diameter, i.e. within a given square (Fig. 2/30)

1. Construct a square PQRS, length of side equal to the diameter.
2. Draw the diagonals SQ and PR to intersect in T.
3. With centres P, Q, R and S draw four arcs, radius PT (= QT = RT = ST) to cut the square in A, B, C, D, E, F, G and H.

ABCDEFGH is the required octagon.

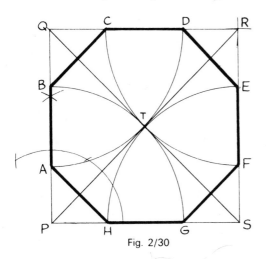

Fig. 2/30

To construct any given polygon, given the length of a side

There are three fairly simple ways of constructing a regular polygon. Two methods require a simple calculation and the third requires very careful construction if it is to be exact. All three methods are shown. The constructions work for any polygon, and a heptagon (seven sides) has been chosen to illustrate them.

Method 1 (Fig. 2/31)

1. Draw a line AB equal in length to one of the sides and produce AB to P.
2. Calculate the exterior angle of the polygon by dividing 360° by the number of sides. In this case the exterior angle is $360°/7 = 51 \ 3°/7$.
3. Draw the exterior angle PBC so that BC = AB.
4. Bisect AB and BC to intersect in O.
5. Draw a circle, centre O and radius OA (= OB = OC).
6. Step off the sides of the figure from C to D, D to E, etc.

ABCDEFG is the required heptagon.

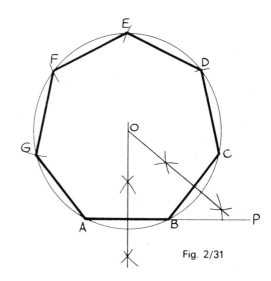

Fig. 2/31

17

Method 2 (Fig. 2/32)

1. Draw a line AB equal in length to one of the sides.
2. From A erect a semi-circle, radius AB to meet BA produced in P.
3. Divide the semi-circle into the same number of equal parts as the proposed polygon has sides. This may be done by trial and error or by calculation ($180°/7 = 25\,5°/7$ for each arc).
4. Draw a line from A to point 2 (for **ALL** polygons). This forms a second side to the polygon.
5. Bisect AB and A2 to intersect in O.
6. With centre O draw a circle, radius OB ($=$ OA $=$ O2).
7. Step off the sides of the figure from B to C, C to D etc.

ABCDEFG is the required septagon.

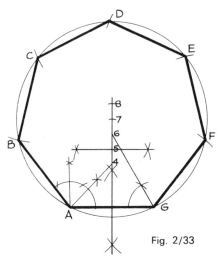

Fig. 2/33

To construct a regular polygon given a diagonal, i.e. within a given circle (Fig. 2/34)

1. Draw the given circle and insert a diameter AM.
2. Divide the diameter into the same number of divisions as the polygon has sides.
3. With centre M draw an arc, radius MA. With centre A draw another arc of the same radius to intersect the first arc in N.
4. Draw N2 and produce to intersect the circle in B (for *any* polygon).
5. AB is the first side of the polygon. Step out the other sides BC, CD, etc.

ABCDE is the required polygon.

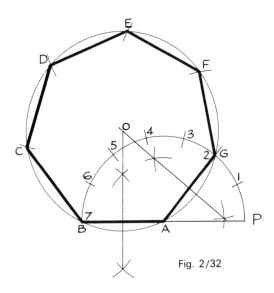

Fig. 2/32

Method 3 (Fig. 2/33)

1. Draw a line GA equal in length to one of the sides.
2. Bisect GA.
3. From A construct an angle of 45° to intersect the bisector at point 4.
4. From G construct an angle of 60° to intersect the bisector at point 6.
5. Bisect between points 4 and 6 to give point 5.

Point 4 is the centre of a circle containing a square. Point 5 is the centre of a circle containing a pentagon. Point 6 is the centre of a circle containing a hexagon. By marking off points at similar distances the centres of circles containing any regular polygon can be obtained.

6. Mark off point 7 so that 6 to 7 = 5 to 6 ($=$ 4 to 5).
7. With centre at point 7 draw a circle, radius 7 to A ($=$ 7 to G).
8. Step off the sides of the figure from A to B, B to C, etc.

ABCDEFG is the required heptagon.

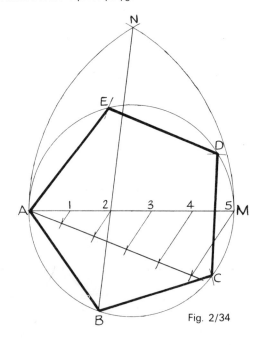

Fig. 2/34

To construct a regular polygon given a diameter
(Fig. 2/35)
1. Draw a line MN.
2. From some point A on the line draw a semi-circle of any convenient radius.

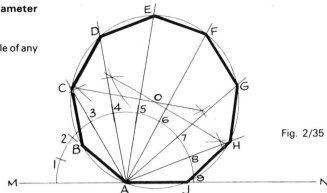

Fig. 2/35

3. Divide the semi-circle into the same number of equal sectors as the polygon has sides (in this case 9, i.e. 20° intervals).
4. From A draw radial lines through points 1 to 8.
5. If the polygon has an even number of sides, there is only one diameter passing through A. In this case, bisect the known diameter to give centre O. If, as in this case, there are two diameters passing through A (there can never be more than two), then bisect both diameters to intersect in O.
6. With centre O and radius OA, draw a circle to intersect the radial lines in C, D, E, F, G and H.
7. From A mark off AB and AJ equal to CD, DE, etc. ABCDEFGHJ is the required polygon.

The constructions shown above are by no means all the constructions that you may be required to do, but they are representative of the type that you may meet.

If your geometry needs a little extra practice, it is well worth while proving these constructions by Euclidean proofs. A knowledge of some geometric theorems is needed when answering many of the questions shown below, and proving the above constructions will make sure that you are familiar with them.

Exercises 2

1. Construct an equilateral triangle with sides 60 mm long.
2. Construct an isosceles triangle that has a perimeter of 135 mm and an altitude of 55 mm.
3. Construct a triangle with base angles 60° and 45° and an altitude of 76 mm.
4. Construct a triangle with a base of 55 mm, an altitude of 62 mm and a vertical angle of $37\frac{1}{2}°$.
5. Construct a triangle with a perimeter measuring 160 mm and sides in the ratio 3:5:6.

6. Construct a triangle with a perimeter of 170 mm and sides in the ratio 7:3:5.
7. Construct a triangle given that the perimeter is 115 mm, the altitude is 40 mm and the vertical angle is 45°.
8. Construct a triangle with a base measuring 62 mm, an altitude of 50 mm and a vertical angle of 60°. Now draw a similar triangle with a perimeter of 250 mm.
9. Construct a triangle with a perimeter of 125 mm, whose sides are in the ratio 2:4:5. Now draw a similar triangle whose perimeter is 170 mm.
10. Construct a square of side 50 mm. Find the mid-point of each side by construction and join up the points with straight lines to produce a second square.
11. Construct a square whose diagonal is 68 mm.
12. Construct a square whose diagonal is 85 mm.
13. Construct a parallelogram given two sides 42 mm and 90 mm long, and the angle between them 67°.
14. Construct a rectangle which has a diagonal 55 mm long and one side 35 mm long.
15. Construct a rhombus if the diagonal is 75 mm long and one side 44 mm long.
16. Construct a trapezium given that the parallel sides are 50 mm and 80 mm long and are 45 mm apart.
17. Construct a regular hexagon, 45 mm side.
18. Construct a regular hexagon if the diameter is 75 mm.
19. Construct a regular hexagon within an 80 mm diameter circle. The corners of the hexagon must all lie on the circumference of the circle.
20. Construct a square, side 100 mm. Within the square, construct a regular octagon. Four alternate sides of the octagon must lie on the sides of the square.
21. Construct the following regular polygons:
    a pentagon, side 65 mm,
    a heptagon, side 55 mm,
    a nonagon, side 45 mm,
    a decagon, side 35 mm.
22. Construct a regular pentagon, diameter 82 mm.
23. Construct a regular heptagon within a circle, radius 60 mm. The corners of the heptagon must lie on the circumference of the circle.

3

Isometric projection

Engineering drawings are always drawn in orthographic projection. For the presentation of detailed drawings, this system has been found to be far superior to all others. The system has, however, the disadvantage of being very difficult to understand by people not trained in its usage. It is always essential that an engineer be able to communicate his ideas to anybody, particularly people who are not engineers, and it is therefore an advantage to be able to draw using a system of projection that is more easily understood. There are many systems of pictorial projection and this book deals with two: isometric and oblique projections. Of these two, isometric presents the more natural looking view of an object.

True isometric projection is an application of orthographic projection and is dealt with in greater detail later

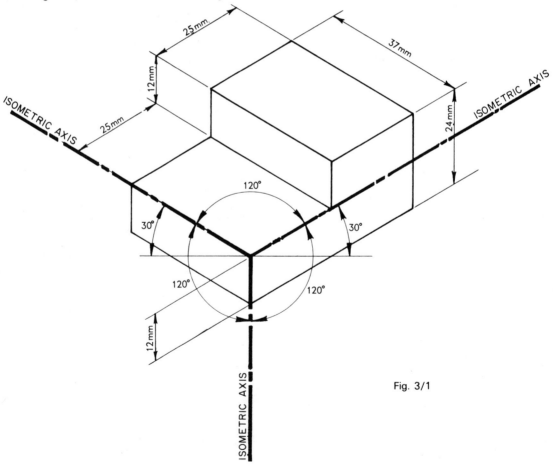

Fig. 3/1

20

in this chapter. The most common form of isometric projection is called 'conventional isometric'. This is the method that is set in G.C.E. 'O' level and C.S.E. examination papers, although a knowledge of true isometric is sometimes assumed.

Conventional isometric projection (isometric drawing)

If you were to make a freehand drawing of a row of houses, the house furthest away from you would be the smallest house on your drawing. This is called the 'perspective' of the drawing and, in a perspective drawing, none of the lines are parallel. Isometric drawing ignores perspective altogether. Lines are drawn parallel to each other and drawings can be made using a tee square and a set square. This is much simpler than perspective drawing.

Fig. 3/1 shows a shaped block drawn in conventional isometric projection.

You will note that there are three isometric axes. These are inclined at 120° to each other. One axis is vertical and the other two axes are therefore at 30° to the horizontal. Dimensions measured along these axes, or parallel to them, are true lengths.

The faces of the shaped block shown in Fig. 3/1 are all at 90° to each other. The result of this is that all of the lines in the isometric drawing are parallel to the isometric axes. If the lines are not parallel to any of the isometric axes, they are no longer true lengths. An example of this is shown in Fig. 3/2 which shows an isometric drawing of a regular hexagonal prism. The hexagon is first drawn as a plane figure and a simple shape, in this case a rectangle, is drawn around the hexagon. The rectangle is easily drawn in isometric and the positions of the corners of the hexagon can be transferred from the plane figure to the isometric drawing with a pair of dividers.

The dimensions of the hexagon should all be 25 mm and you can see from Fig. 3/2 that lines not parallel to the isometric axes do not have true lengths.

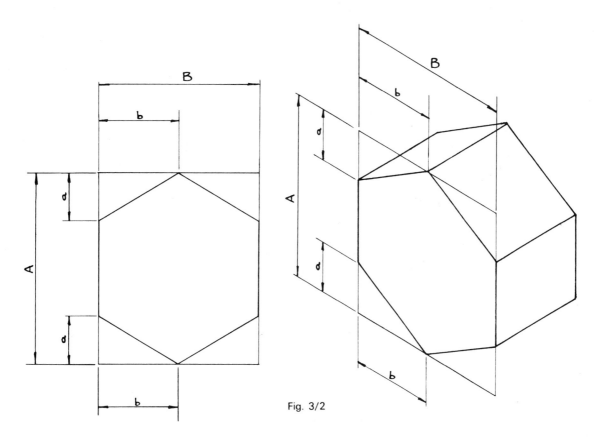

Fig. 3/2

Fig. 3/3 shows another hexagonal prism. This prism has been cut at an incline and this means that two extra views must be drawn so that sufficient information to draw the prism in isometric can be transferred from the plane views to the isometric drawing.

This figure shows that, when making an isometric drawing, all dimensions must be measured parallel to one of the isometric axes.

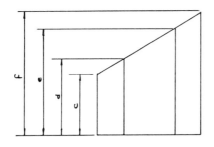

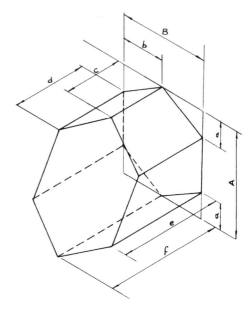

Fig. 3/3

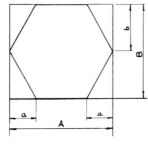

Circles and curves drawn in isometric projection

All of the faces of a cube are square. If a cube is drawn in isometric projection, each square side becomes a rhombus. If a circle is drawn on the face of a cube, the circle will change shape when the cube is drawn in isometric projection. Fig. 3/4 shows how to plot the new shape of the circle.

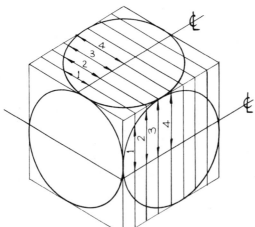

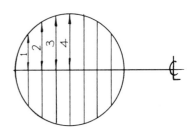

Fig. 3/4

The circle is first drawn as a plane figure, and is then divided into an even number of equal strips. The face of the cube is then divided into the same number of equal strips. Centre lines are added and the measurement from the centre line of the circle to the point where strip 1 crosses the circle is transferred from the plane drawing to the isometric drawing with a pair of dividers. This measurement is applied above and below the centre line. This process is repeated for strips 2, 3, etc.

The points which have been plotted should then be carefully joined together with a neat freehand curve.

Fig. 3/5 illustrates how this system is used in practice.

Since a circle can be divided into four symmetrical quadrants, it is really necessary to draw only a quarter of a circle instead of a whole plane circle.

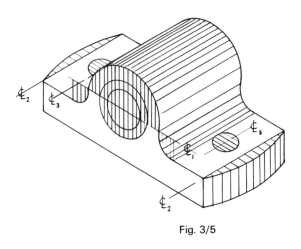

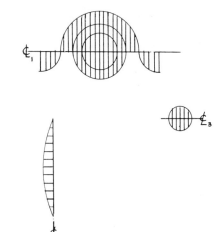

Fig. 3/5

The dimensions which are transferred from the plane circle to the isometric view are called ordinates and the system of transferring ordinates from plane figures to isometric views is not confined to circles. It may be used for any regular or irregular shape. Fig. 3/6 shows a shaped plate.

There are several points worth noting from Fig. 3/6.

(a) Since the plate is symmetrical about its centre line, only half has been divided into strips on the plane figure.

(b) In proportion to the plate, the holes are small. They have, therefore, ordinates much closer together so that they can be drawn accurately.

(c) The point where the vee cut-out meets the elliptical outline has its own ordinate so that this point can be transferred exactly to the isometric view.

(d) Since the plate has a constant thickness, the top and bottom profiles are the same. A quick way of plotting the bottom profile is to draw a number of vertical lines down from the top profile and, with dividers set at the required thickness of plate, follow the top curve with the dividers, marking the thickness of the plate on each vertical line.

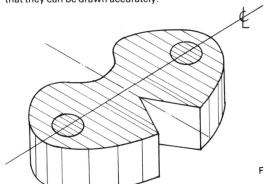

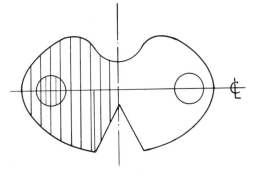

Fig. 3/6

It is sometimes necessary to draw circles or curves on faces which are not parallel to any of the three isometric axes. Fig. 3/7 shows a cylinder cut at 45°. Two views of the cylinder have to be drawn: a plan view and an elevation. The plan view is divided into strips and the positions of these strips are projected onto the elevation.

The base of the cylinder is drawn in isometric in the usual way. Points 1 to 20, where the strips cross the circle, are projected vertically upwards and the height of the cylinder, measured from the base with dividers, is transferred for each point in turn from the elevation to the isometric view. These points are then carefully joined together with a neat freehand curve.

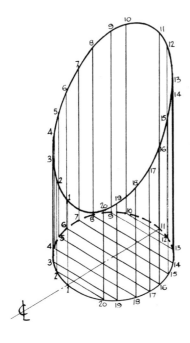

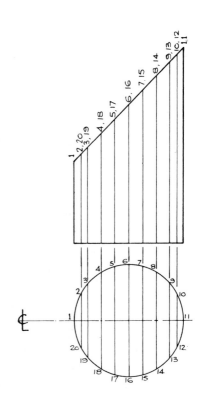

Fig. 3/7

True isometric projection
Isometric projection is a method of drawing with instruments which gives a pictorial view of an object. It is not often used in industry and, when it is used, the vast majority of drawings would be made using conventional isometric projection. Conventional isometric is a distorted and simplified form of true isometric. True isometric is found by taking a particular view from an orthographic

projection of an object. Fig. 3/8 shows a cube, about 25 mm side, drawn in orthographic projection with the cube so positioned that the front elevation is a true isometric projection of the cube. The three isometric axes are still at 120° to each other. In conventional isometric, distances measured parallel to these axes are true lengths. In true isometric projection they are no longer true lengths although they are proportional to their true lengths. How-

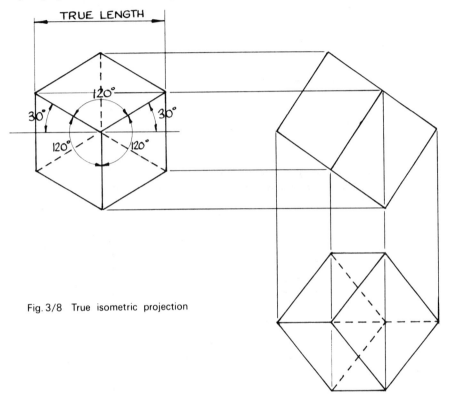

Fig. 3/8 True isometric projection

ever, the horizontal distances on a true isometric projection *are* true lengths. The reduction of lengths measured parallel to the isometric axes makes the overall size of the true isometric drawing appear to look more natural, particularly when directly compared with an orthographic or plane view of the same object (compare the relative sizes of the prism in Fig. 3/2).

If the horizontal length and the length parallel to the isometric axes were both to be true lengths, the isometric axes would have to be at 45° (Fig. 3/9). Since the isometric axes are at 30°, the 45° lengths must be reduced.

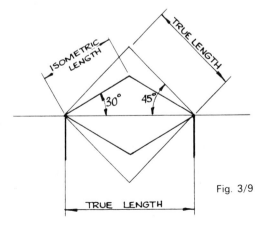

Fig. 3/9

This operation is shown in Fig. 3/10.

The ratio between the true length and the isometric length is ISOMETRIC LENGTH = TRUE LENGTH × 0·8165.

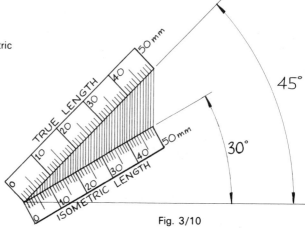

Fig. 3/10

This ratio is constant for all lines measured parallel to any of the isometric axes. If you are asked to draw an object using an isometric scale, your scale may be constructed as in Fig. 3/10 or you may construct a conventional plain scale as shown in Fig. 3/11. The initial length of this scale is 100 × 0.8165 = 81.65 mm. The scale is then completed as shown in Chapter 1.

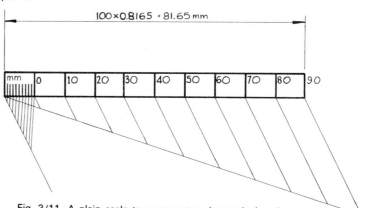

Fig. 3/11 A plain scale to measure true isometric lengths

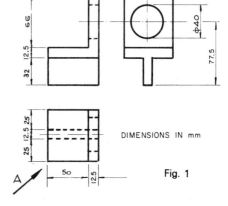

Exercises 3

(All questions originally set in Imperial units)
1. Draw full size an isometric projection of the component shown in Fig. 1 looking in the direction of the arrow A. Hidden details are not to be shown.
Associated Lancashire Schools Examining Board

DIMENSIONS IN mm

Fig. 1

2. Fig. 2 shows the front elevation and plan of an ink bottle stand. Make a full size isometric drawing of the stand with corner A nearest to you. Hidden details should *not* be shown.
West Midlands Examinations Board

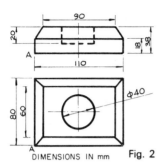

DIMENSIONS IN mm   **Fig. 2**

3. Fig. 3 shows the development of a hexagonal box. Draw, in isometric projection, the assembled box standing on its base. Ignore the thickness of the material and omit hidden detail.
North Western Secondary School Examinations Board (See Ch. 14 for information not in Ch. 3).

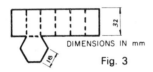

DIMENSIONS IN mm

Fig. 3

4. Three views of a bearing are shown in Fig. 4. Make an isometric drawing of the bearing. Corner A should be the lowest point on your drawing. No hidden details are required.
South-East Regional Examinations Board

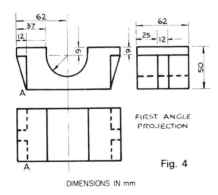

FIRST ANGLE
PROJECTION

Fig. 4

DIMENSIONS IN mm

5. Two views of a plain shaft bearing are shown in Fig. 5. Make a full size isometric drawing of the bearing. Hidden details should *not* be shown.
West Midlands Examinations Board

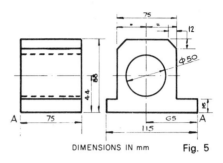

DIMENSIONS IN mm   **Fig. 5**

6. Construct an isometric drawing of the casting shown in Fig. 6. Make point X the lowest point of your drawing. Do not use an isometric scale.
University of London School Examinations

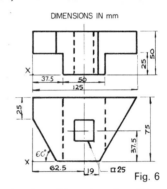

Fig. 6

7. Construct an isometric scale and use it to make a true isometric view of the casting shown in Fig. 7. Corner X is to be the lowest corner of your drawing.
University of London School Examinations

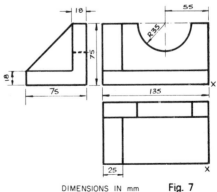

DIMENSIONS IN mm   **Fig. 7**

8. A plan and elevation of the base of a candlestick are shown in Fig. 8. Draw (a) another elevation when the base is viewed in the direction of the arrow and (b) an accurate isometric view of that half of the candlestick base indicated by the letters *abcd* on the plan view. The edge *ab* is to be in the foreground of your drawing.

   Southern Universities' Joint Board (See Ch. 8 for information not in Ch. 3).

10. Fig. 10 shows two views of a cylindrical rod with a circular hole. Make an isometric drawing of the rod in the direction of the arrows R and S. Hidden detail is not required.
    Associated Examining Board

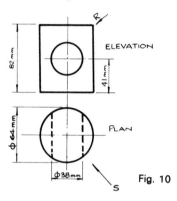

Fig. 10

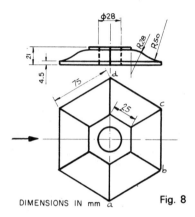

DIMENSIONS IN mm   **Fig. 8**

9. Fig. 9 shows the plan and elevation of an angle block. Make a full size isometric projection of the block, making the radiused corner the lowest part of your drawing. Do *not* use an isometric scale. Hidden detail should be shown.
   Oxford and Cambridge Schools Examinations Board

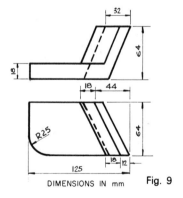

DIMENSIONS IN mm   **Fig. 9**

28

4

The construction of circles to satisfy given conditions

About 6000 years ago, an unknown Mesopotamian made one of the greatest inventions of all time, the wheel. This was the most important practical application ever made of a shape that fascinated early mathematicians. The shape is, of course, the circle. After the wheel had been invented, the Mesopotamians found many more applications for the circle than just for transport. The potter's wheel was developed and vessels were made much more accurately and quickly. Pulleys were invented and engineers and builders were able to raise heavy weights. Since that time, the circle has been the most important geometric shape in the development of all forms of engineering.

Apart from its practical applications, the circle has an aesthetic value which makes it unique amongst plane figures. The ancients called it 'the perfect curve' and its symmetry and simplicity has led artists and craftsmen to use the circle as a basis for design for many thousands of years.

Definitions
A circle is the locus of a point which moves so that it is always a fixed distance from another stationary point.
  Concentric circles are circles that have the same centre.
  Eccentric circles are circles that are not concentric.
  Fig. 4/1 shows some of the parts of the circle.

Constructions
The length of the circumference of a circle is ΠD or $2\Pi R$, where D is the diameter and R the radius of the circle. Π is the ratio of the diameter to the circumference and may be taken as 22/7 or, more accurately, as 3.142.

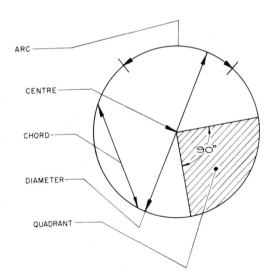

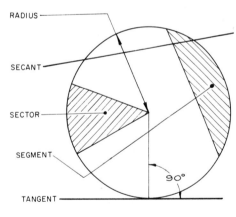

Fig. 4/1

If you need to draw the circumference of a circle (this is required quite often in subsequent chapters), you should either calculate it, or use the construction shown in Fig. 4/2. This construction is not exact but is accurate enough for most needs. For the sake of thoroughness, the corresponding construction, that of finding the diameter from the circumference, is shown in Fig. 4/3.

To construct the circumference of a circle, given the diameter (Fig. 4/2)
1. Draw a semi-circle of the given diameter AB, centre O.
2. From B mark off three times the diameter, BC.
3. From O draw a line at 30° to OA to meet the semi-circle in D.
4. From D draw a line perpendicular to OA to meet OA in E.
5. Join EC.
EC is the required circumference.

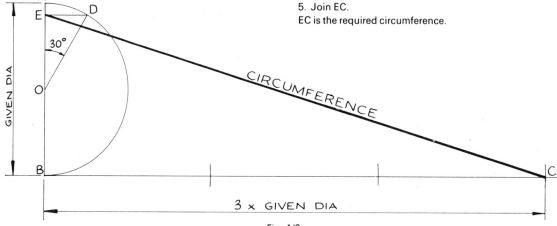

Fig. 4/2

To construct the diameter of a circle, given the circumference (Fig. 4/3)

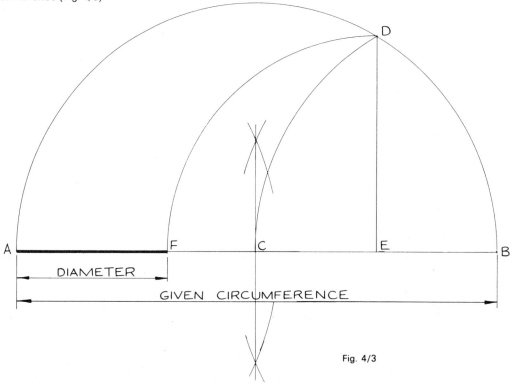

Fig. 4/3

1. Draw the given circumference AB.
2. Bisect AB in C.
3. With centre C, and radius CA, draw a semi-circle.
4. With centre B, and radius BC, draw an arc to cut the semi-circle in D.
5. From D draw a perpendicular to AB, to cut AB in E.
6. With centre E and radius ED draw an arc to cut AB in F.

AF is the required diameter.

The rest of this chapter shows some of the constructions for finding circles drawn to satisfy certain given conditions.

To find the centre of any circle (Fig. 4/4)
1. Draw any two chords.
2. Construct perpendicular bisectors to these chords to intersect in O.

O is the centre of the circle.

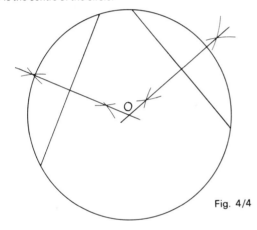

Fig. 4/4

To construct a circle to pass through three given points (Fig. 4/5)
1. Draw straight lines connecting the points as shown. These lines are, in fact, chords of the circle.

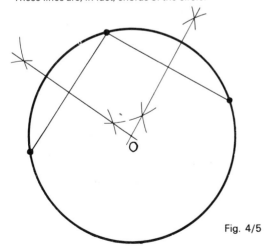

Fig. 4/5

2. Draw perpendicular bisectors through these lines to intersect in O.

O is the centre of a circle which passes through all three points.

To construct the inscribed circle of any regular polygon (in this case, a triangle) (Fig. 4/6)
1. Bisect any two of the interior angles to intersect in O. (If the third angle is bisected it should also pass through O.)

O is the centre of the inscribed circle. This centre is called the incentre.

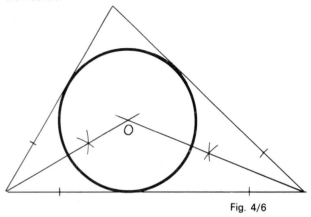

Fig. 4/6

To construct the circumscribed circle of any regular polygon (in this case a triangle) (Fig. 4/7)
1. Perpendicularly bisect any two sides to intersect in O. (If the third side is bisected it should also pass through O.)

O is the centre of the circumscribed circle. This centre is called the circumcentre.

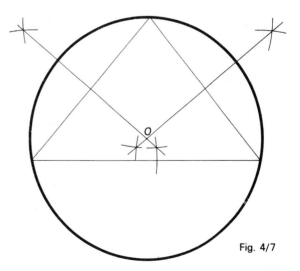

Fig. 4/7

To construct the escribed circle to any regular polygon (in this case a triangle) (Fig. 4/8)

1. An escribed circle is a circle which touches a side and the two adjacent sides produced. Thus, the first step is to produce the adjacent sides.
2. Bisect the exterior angles thus formed to intersect in O.

O is the centre of the escribed circle.

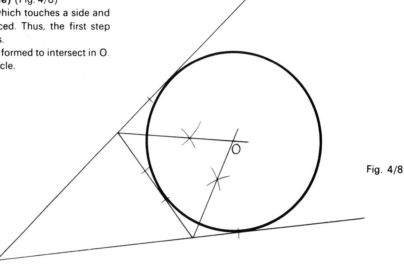

Fig. 4/8

To construct a circle which passes through a fixed point A and touches a line at a given point B (Fig. 4/9)

1. Join AB.
2. From B erect a perpendicular BC.
3. From A construct angle BÂO similar to angle CB̂A to intersect the perpendicular in O.

O is the centre of the required circle.

Fig. 4/9

To construct a circle which passes through two given points, A and B, and touches a given line (Fig. 4/10)

1. Join AB and produce this line to D (cutting the given line in C) so that BC = CD.
2. Construct a semi-circle on AD.
3. Erect a perpendicular from C to cut the semi-circle in E.

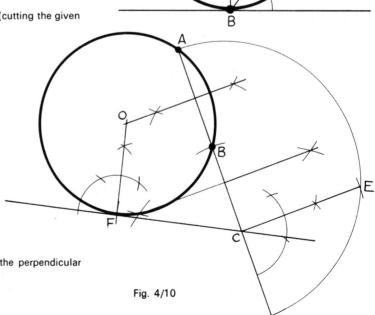

4. Make CF = CE.
5. From F erect a perpendicular.
6. Perpendicularly bisect AB to meet the perpendicular from F in O.

O is the centre of the required circle.

Fig. 4/10

To construct a circle which touches two given lines and passes through a given point P. (There are two circles which satisfy these conditions (Fig. 4/11)

1. If the two lines do not meet, produce them to intersect in A.
2. Bisect the angle thus formed.
3. From any point on the bisector draw a circle, centre B, to touch the two given lines.
4. Join PA to cut the circle in C and D.
5. Draw PO_1 parallel to CB and PO_2 parallel to DB. O_1 and O_2 are the centres of the required circles.

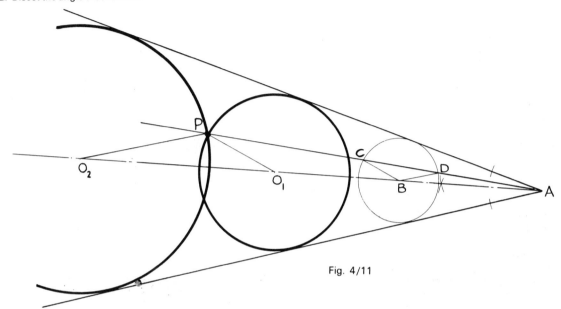

Fig. 4/11

To construct a circle, radius R, to touch another given circle radius r, and a given line (Fig. 4/12)

1. Draw a line parallel to the given line, the distance between the lines equal to R.
2. With compass point at the centre of the given circle and radius set at $R + r$, draw an arc to cut the parallel line in O.

O is the centre of the required circle.

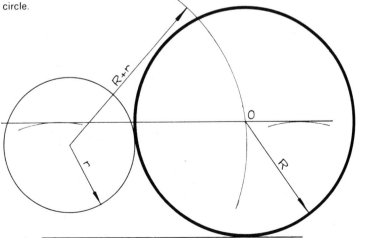

Fig. 4/12

To construct a circle which touches another circle and two tangents of that circle (Fig. 4/13)

1. If the tangents do not intersect, produce them to intersect in A.
2. Bisect the angle formed by the tangents.
3. From B, the point of contact of the circle and one of its tangents, construct a perpendicular to cut the bisector in O_1. This is the centre of the given circle.
4. Join BD.
5. Draw EF parallel to DB and FO_2 parallel to BO_1. O_2 is the centre of the required circle.

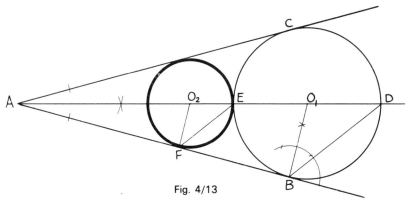

Fig. 4/13

To construct a circle which touches another circle and two lines (Fig. 4/14)

1. Draw intersecting lines parallel to the given lines. These lines, AB and AC, must be distance r, the radius of the given circle, from the given lines.
2. Repeat construction 4/11 and construct a circle which passes through O_1, the centre of the given circle, and touches the two parallel lines.
3. The centre of this circle, O_2, is also the centre of the required circle.

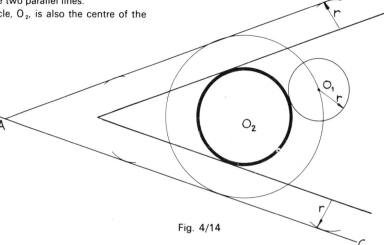

Fig. 4/14

To construct a circle which passes through two given points, P and Q, and touches a given circle, centre D (Fig. 4/15)

1. Join PQ and produce this line.
2. Perpendicularly bisect PQ and, with centre somewhere on this bisector, draw a circle to pass through points P and Q and cut the given circle in A and B.
3. Join AB and produce to cut PQ produced in C.
4. Construct the tangent from C to the given circle. (Join CD, bisect CD in E, compass point at E draw a radius ED to cut the circle in F).
5. From F erect a perpendicular to cut the bisector of PQ in O.

O is the centre of the required circle.

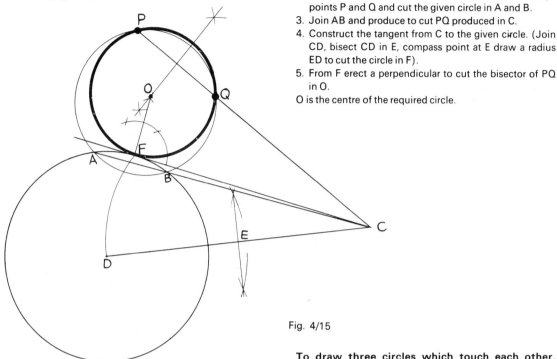

Fig. 4/15

All of the above constructions are for finding single circles which satisfy given conditions. The rest of the constructions in this chapter are concerned with more than one circle at a time.

To draw three circles which touch each other, given the position of their centres O_1, O_2 and O_3 (Fig. 4/16)

1. Draw straight lines connecting the centres.
2. Find the centre of the triangle thus formed by bisecting two of the interior angles.
3. From this centre, drop a perpendicular to cut $O_1 O_2$ in A.
4. With centre O_1 and radius O_1A, draw the first circle.
5. With centre O_2 and radius O_2A, draw the second circle.
6. With centre O_3 and radius O_3C ($= O_3B$), draw the third circle.

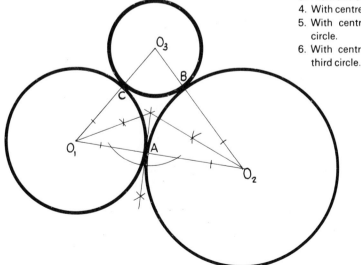

Fig. 4/16

To draw two circles, given both their radii, within a third circle, all three circles to touch each other (Fig. 4/17)

1. Mark off the diameter AB of the largest circle.
2. Mark off AO₁, equal to the radius of one of the other circles and draw this circle, centre O₁, to cut the diameter in C.
3. From C mark off CD equal to the radius of the third circle.
4. Mark off BE equal to the radius of the third circle.
5. With centre O₁ and radius O₁D, draw an arc.
6. With centre O and radius OE, draw an arc to cut the first arc in O₂.

O₂ is the centre of the third circle.

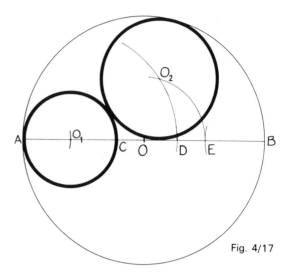

Fig. 4/17

To draw any number of equal circles within another circle, the circles all to be in contact (in this case 5) (Fig. 4/18)

1. Divide the circle into the same number of sectors as there are proposed circles.
2. Bisect all the sectors and produce one of the bisectors to cut the circle in D.
3. From D erect a perpendicular to meet OB produced in E.
4. Bisect DÊO to meet OD in F.
5. F is the centre of the first circle. The other circles have the same radius and have centres on the intersections of the sector bisectors and a circle, centre O and radius OF.

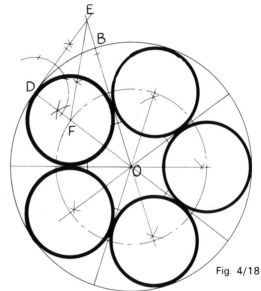

Fig. 4/18

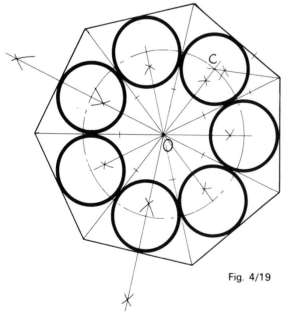

Fig. 4/19

To draw a number of equal circles within a regular polygon to touch each other and one side of the polygon (in this case, a septagon) (Fig. 4/19)

1. Find the centre of the polygon by bisecting two of the sides.
2. From this centre, draw lines to all of the corners.
3. This produces a number of congruent triangles. All we now need to do is to draw the inscribed circle in each of these triangles. This is done by bisecting any two of the interior angles to give the centre C.
4. The circles have equal radii and their centres lie on the intersection of a circle, radius OC and the bisectors of the seven equal angles formed by step 2.

To draw equal circles around a regular polygon to touch each other and one side of the polygon (in this case, a septagon) (Fig. 4/20)

1. Find the centre of the polygon by bisecting two of the sides.
2. From the centre O draw lines through all of the corners and produce them.

3. Bisect angles CÂB and DB̂A to intersect in E.
4. E is the centre of the first circle. The rest can be obtained by drawing a circle, radius OE, and bisecting the seven angles formed by step 2. The intersections of this circle and these lines give the centres of the other six circles.

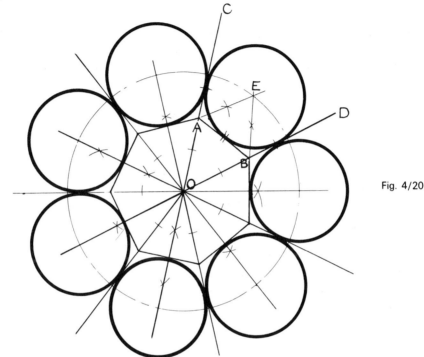

Fig. 4/20

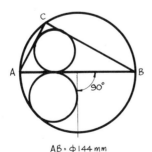

AB = φ144 mm
AC = 68 mm

Fig. 1

Exercises 4

(All questions originally set in Imperial units)

1. Construct a regular octagon on a base line 25 mm long and draw the inscribed circle. Measure and state the diameter of this circle in mm.
   North Western Secondary School Examinations Board (See Ch. 2 for information not in Ch. 4)
2. Describe *three* circles, each one touching the other two externally, their radii being 12 mm, 18 mm and 24 mm respectively.
   North Western Secondary School Examinations Board
3. No construction has been shown in Fig. 1. You are required to draw the figure full size showing all construction lines necessary to ensure the circles are tangential to their adjacent lines.
   Southern Regional Examinations Board

4. Construct the triangle ABC in which the base BC = 108 mm, the vertical angle Â = 70° and the altitude is 65 mm.
D is a point on AB 34 mm from A. Describe a circle to pass through the points A and D and touch (tangential to) the line BC.
Southern Universities' Joint Board (See Ch. 2 for information not in Ch. 4)

5. Fig. 2 shows two touching circles placed in the corner made by two lines which are perpendicular to one another. Draw the view shown and state the diameter of the smaller circle. Your construction must show clearly the method of obtaining the centre of the smaller circle.
University of London School Examinations

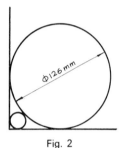

Fig. 2

6. Fig. 3 shows two intersecting lines AB and BC and the position of a point P. Draw the given figure and find the centre of a circle which will pass through P and touch the lines AB and BC. Draw the circle and state its radius as accurately as possible.
University of London School Examinations

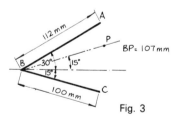

Fig. 3

7. A triangle has sides 100 mm, 106 mm and 60 mm long. Draw the triangle and construct and draw the following: (a) the inscribed circle; (b) the circumscribed circle; (c) the smallest escribed circle.
University of London School Examinations (See Ch. 2 for information not in Ch. 4)

8. Construct an isosceles triangle ABC where the included angle A = 67½°, and AB = AC = 104 mm.
Draw circles of 43 mm, 37 mm and 32 mm radius using as centres A, B and C respectively.
Construct the smallest circle which touches all three circles.
Measure and state the diameter of the constructed circle.
Associated Examining Board (See Ch. 2 for information not in Ch. 4)

9. AB and AC are two straight lines which intersect at an angle of 30°. D is a point between the two lines at perpendicular distances of 37 mm and 62 mm respectively from AB and AC. Describe the circle which touches the two converging lines and passes through point D; the centre of this circle is to lie between the points A and D. Now draw two other circles each touching the constructed circle externally and also the converging lines. Measure and state the diameters of the constructed circles.
Oxford Local Examinations

10. OA and OB are two straight lines meeting at an angle of 30°. Construct a circle of diameter 76 mm to touch these two lines and a smaller circle which will touch the two converging lines and the first circle.
Also construct a third circle of diameter 64 mm which touches each of the other two circles.
Oxford Local Examinations

11. Construct a regular octagon of side 75 mm and within this octagon describe eight equal circles each touching one side of the octagon and two adjacent circles. Now draw the smallest circle which will touch all eight circles. Measure and state the diameter of this circle.
Oxford Local Examinations (See Ch. 2 for information not in Ch. 4)

5
Tangency

Definition
A tangent to a circle is a straight line which touches the circle at one point.

Every curve ever drawn could have tangents drawn to it, but this chapter is concerned only with tangents to circles. These have wide applications in Engineering Drawing since the outlines of most engineering details are made up of straight lines and arcs. Wherever a straight line meets an arc, a tangent meets a circle.

Constructions
To draw a tangent to a circle from any point on the circumference (Fig. 5/1)
1. Draw the radius of the circle.
2. At any point on the circumference of a circle, the tangent and the radius are perpendicular to each other.
   Thus, the tangent is found by constructing an angle of 90° from the point where the radius crosses the circumference.

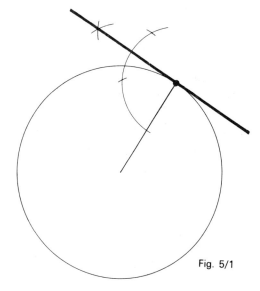

Fig. 5/1

A basic geometric theorem is that the angle in a semi-circle is a right angle (Fig. 5/2).
This fact is made use of in many tangent constructions.

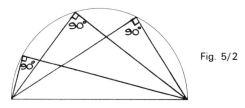

Fig. 5/2

To construct a tangent from a point P to a circle, centre O (Fig. 5/3)

1. Join OP.
2. Erect a semi-circle on OP to cut the circle in A.

PA produced is the required tangent (OA is the radius and is perpendicular to PA since it is the angle in a semi-circle). There are, of course, two tangents to the circle from P but only one has been shown for clarity.

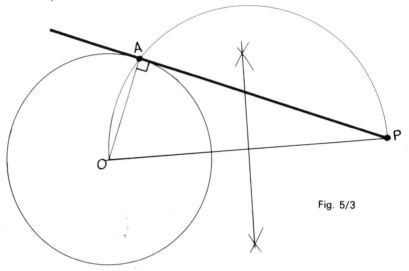

Fig. 5/3

To construct a common tangent to two equal circles (Fig. 5/4)

1. Join the centres of the two circles.
2. From each centre, construct lines at 90° to the centre line. The intersection of these perpendiculars with the circles gives the points of tangency.

This tangent is often described as the common exterior tangent.

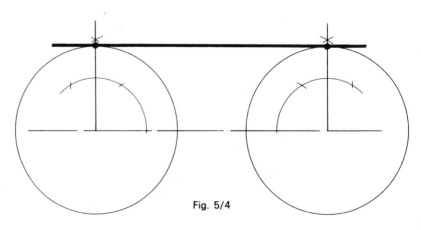

Fig. 5/4

To construct the common interior (or transverse or cross) tangent to two equal circles, centres O and O₁ (Fig. 5/5)

1. Join the centres O O₁.
2. Bisect O O₁ in A.
3. Bisect OA in B and draw a semi-circle, radius BA to cut the circle in C.
4. With centre A and radius AC draw an arc to cut the second circle in D.

CD is the required tangent.

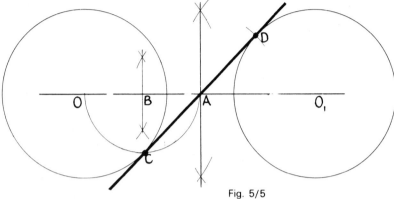

Fig. 5/5

To construct the common tangent between two unequal circles, centres 0 and O₁ and radii R and r respectively (Fig. 5/6)

1. Join the centres O O₁.
2. Bisect O O₁ in A and draw a semi-circle, radius AO.
3. Draw a circle, centre O, radius R-r, to cut the semi-circle in B.
4. Join OB and produce to cut the larger circle in C.
5. Draw O₁D parallel to OC.

CD is the required tangent.

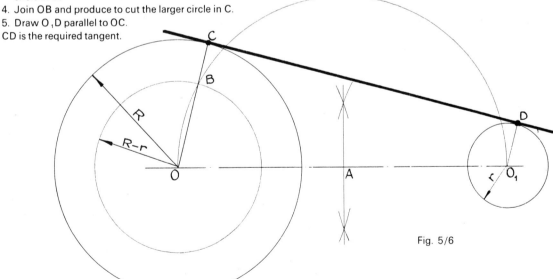

Fig. 5/6

41

To construct the common internal tangent between two unequal circles, centres O and O_1, and radii R and r respectively (Fig. 5/7)

1. Join the centres $O O_1$.
2. Bisect $O O_1$ in A and draw a semi-circle, radius OA.
3. Draw a circle, centre O, radius $R + r$, to cut the semi-circle in B.
4. Join OB. This cuts the larger circle in C.
5. Draw O_1D parallel to OB.

CD is the required tangent.

A tangent is, by definition, a straight line. However, we do often talk of radii or curves meeting each other tangentially. We mean, of course, that the curves meet smoothly and with no change of shape or bumps. This topic, the blending of lines and curves, is discussed in Chapter 8.

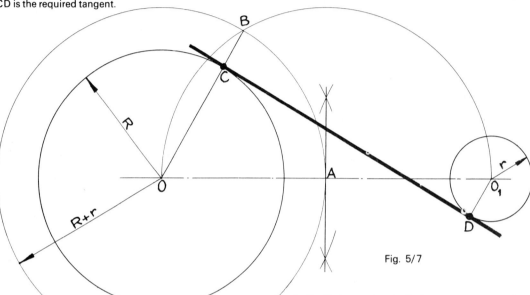

Fig. 5/7

Exercises 5

(All questions originally set in Imperial units)

1. A former in a jig for bending metal is shown in Fig. 1.
   (a) Draw the former, full size, showing in full the construction for obtaining the tangent joining the two arcs.
   (b) Determine, without calculation, the centres of the four equally spaced holes to be bored in the positions indicated in the figure.

Middlesex Regional Examining Board

2. Fig. 2 shows a centre finder, or centre square in position on a 75 mm diameter bar.
   Draw, *full size*, the shape of the centre finder and the piece of round bar. Show clearly the constructions for
   (a) the tangent, AA, to the two arcs;
   (b) the points of contact and the centre for the 44 mm rad at B;
   (c) the points of contact and the centre for the 50 mm rad at C.

South-East Regional Examinations Board (See Ch. 8 for information not in Ch. 5)

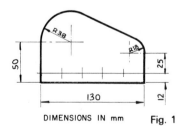

DIMENSIONS IN mm   **Fig. 1**

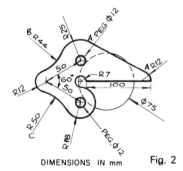

DIMENSIONS IN mm   **Fig. 2**

3. Fig. 3 shows the outline of two pulley wheels connected by a belt of negligible thickness. To a scale of 1/10 draw the figure showing the construction necessary to obtain the points of contact of the belt and pulleys.
Middlesex Regional Examining Board

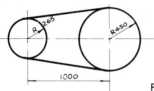

Fig. 3

4. (1) Draw the figure ABCP shown in Fig. 4 and construct a circle, centre O, to pass through the points A, B and C.

(2) Construct a tangent to this circle touching the circle at point B.

(3) Construct a tangent from the point P to touch the circle on the minor arc of the chord AC.

Southern Regional Examining Board (See Ch. 4 for information not in Ch. 5)

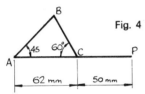

Fig. 4

5. Fig. 5 shows a metal blank. Draw the blank, full size, showing clearly the constructions for obtaining the tangents joining the arcs.

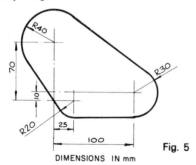

Fig. 5

DIMENSIONS IN mm

Fig. 6 shows the outlines of three pulley wheels connected by a taut belt. Draw the figure, full size, showing clearly the constructions for obtaining the points of contact of the belt and pulleys.

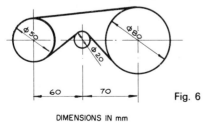

Fig. 6

DIMENSIONS IN mm

7. Fig. 7 shows the outline of a metal blank. Draw the blank, full size, showing clearly the constructions for finding exact positions of the tangents joining the arcs.

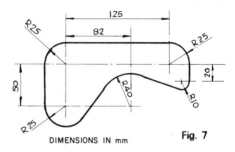

DIMENSIONS IN mm   Fig. 7

8. A segment of a circle stands on a chord AB which measures 50 mm. The angle in the segment is 55°. Draw the segment. Produce the chord AB to C making BC 56 mm long. From C construct a tangent to the arc of the segment.
University of London School Examinations (See Ch. 2 and Ch. 4 for information not in Ch. 5)

9. A and B are two points 100 mm apart. With B as centre draw a circle 75 mm diameter. From A draw two lines AC and AD which are tangential to the circle AC = 150 mm. From C construct another tangent to the circle to form a triangle ACD. Measure and state the lengths CD and AD, also angle CDA.
Joint Matriculation Board

10. Fig. 8 shows two circles, A and B, and a common external tangent and a common internal tangent. Construct (a) the given circles and tangents and (b) the smaller circle which is tangential to circle B and the two given tangents.

Measure and state the distance between the centres of the constructed circle and circle A.

Associated Examining Board (See Ch. 4 for information not in Ch. 5)

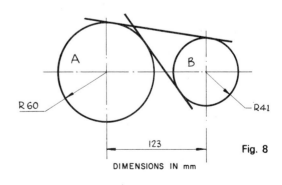

Fig. 8

DIMENSIONS IN mm

43

6

Oblique projection

Oblique projection is another method of pictorial drawing. It is simpler than isometric but it does not present so realistic a picture.

Fig. 6/1 shows a shaped block drawn in oblique projection.

There are three drawings of the same block in Fig. 6/1. They all show the front face of the block drawn in the plane of the paper and the side and top faces receding at 30°, 45° and 60° on the three drawings. An oblique line is one which is neither vertical nor horizontal, and the receding lines in oblique projection can be at any angle other than 0° or 90° as long as they remain parallel in any one drawing. In practice, it is usual to keep to the set square angles and, of the three to choose from, 45° is the most widely used.

If you check the measurements on the oblique drawings with those on the isometric sketch, you will find that the measurements on the front and oblique faces are all true lengths. This gives rise to a distorted effect. The drawings of the block in the oblique view appear to be out of proportion, particularly when compared with the isometric view.

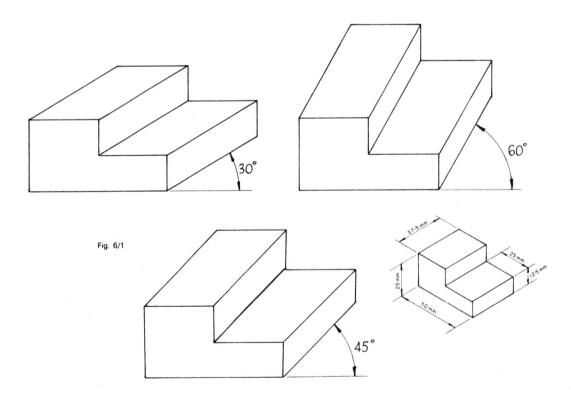

Fig. 6/1

Fig. 6/2 shows how we attempt to overcome this distortion.

The oblique lengths have been altered. The degree of alteration has been determined by the oblique angle. An oblique angle of 60° causes a large distortion and the oblique length is thus altered to $\frac{1}{3}$ × the true length. 30° causes less distortion and the oblique length is only altered to $\frac{3}{4}$ × the true length. At 45° the true length is reduced by half. These alterations need not be rigidly adhered to. The ones illustrated are chosen because they produce a reasonably true to life picture of the block, but a complicated component might have to be drawn with no reduction at all in order to show all the details clearly.

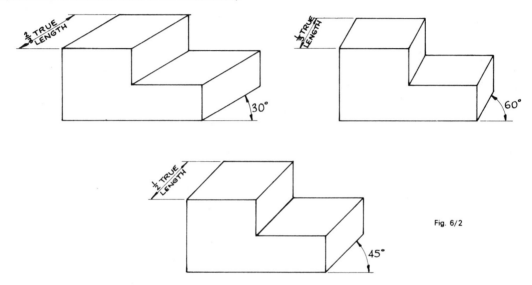

Fig. 6/2

If an oblique drawing is made without any reduction in oblique length, this is sometimes known as Cavalier Projection. If a reduction in oblique length is made, this is sometimes known as Cabinet Projection.

If you were now asked to draw an object in oblique projection, you would probably be very confused when trying to decide which angle to choose and what reduction to make on the oblique lines. If you are asked to produce an oblique drawing, *draw at an oblique angle of 45° and reduce all your oblique dimensions by half, unless you are given other specific instructions.*

Circles and curves in Oblique Projection

Oblique projection has one very big advantage over isometric projection. Since the front face is drawn in the plane of the paper, any circles on this face are true circles and not ellipses as was the case with isometric projection. Fig. 6/3 shows an oblique drawing of a bolt. If the bolt had been drawn in isometric, it would have been a long and tedious drawing to make.

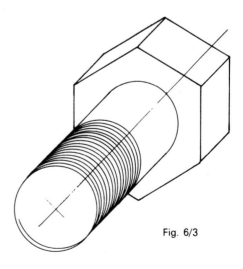

Fig. 6/3

There are occasions when there are curves or circles on the oblique faces. When this arises, they may be drawn using the ordinate method that was used for circles on isometric drawings. If the oblique length has been scaled down, then the ordinates on the oblique lengths must be scaled down in the same proportions. Fig. 6/4 shows an example of this.

In this case, the oblique angle is 45° and the oblique scale is $\frac{1}{2}$ normal size. The normal 6 mm ordinates are reduced to 3 mm on the oblique faces and the 3 mm ordinates are reduced to 1.5 mm.

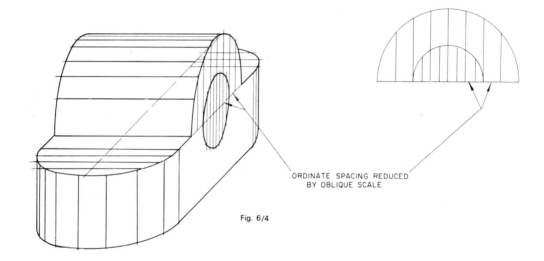

ORDINATE SPACING REDUCED
BY OBLIQUE SCALE

Fig. 6/4

It is also worth noting that the ordinates are spaced along a 45° line. This must always be done in oblique projection in order to scale the distances between the ordinates on the oblique view to half those on the plane view.

The advantage of oblique projection over other pictorial projections is that circles drawn on the front face are not distorted. Unfortunately, examiners usually insist that circles are drawn on the oblique faces, as in Fig. 6/4. However, if you are free of the influence of an examiner and wish to draw a component in oblique projection, it is obviously good sense to ensure that the face with the most circles or curves is the front face.

Fig. 6/5 shows a small stepped pulley drawn twice in oblique projection. It is obvious that the drawing on the left is easier to draw than the one on the right.

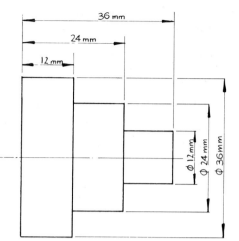

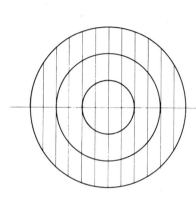

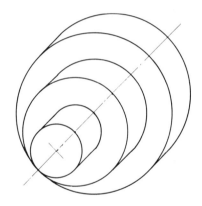

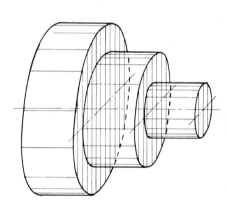

Fig. 6/5

Exercises 6

(All questions originally set in Imperial units)

1. Fig. 1 shows two views of a small casting. Draw, full size, an oblique projection of the casting with face A towards you.

2. Fig. 2 shows two views of a cast iron hinge block. Make an oblique drawing of this object, with face A towards you, omitting hidden detail.
North Western Secondary School Examinations Board

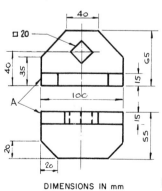

DIMENSIONS IN mm   Fig. 1

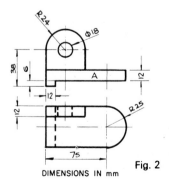

DIMENSIONS IN mm   Fig. 2

3. Fig. 3 shows the outline of the body of a depth gauge. Make an oblique drawing, twice full size, of the body with corner A towards you.

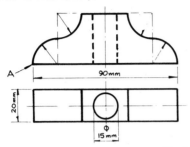

ALL RADII = 15 mm   Fig. 3

4. Draw, full size, an Oblique Projection of the car brake light switch shown in Fig. 4. It should be positioned so that it is resting on surface A, with the cylinder B towards you.
South-East Regional Examinations Board

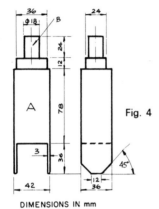

Fig. 4

DIMENSIONS IN mm

5. Fig. 5 shows two views of a holding-down clamp. Draw the clamp, full size, in oblique projection with corner A towards you.

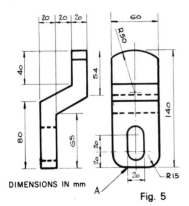

DIMENSIONS IN mm

Fig. 5

6. Two views of a casting are shown in Fig. 6. Draw (a) the given views, and (b) an oblique pictorial view, looking in the direction shown by arrow L, using cabinet projection, that is with the third dimension at 45° and drawn half-size.
Associated Examining Board

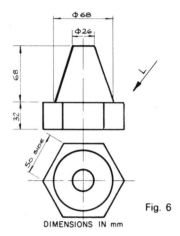

Fig. 6

DIMENSIONS IN mm

7. Two views of a machined block are shown in Fig. 7. Draw, full size, an oblique projection of the block with AB sloping upward to the right at an angle of 30°. Use half-size measurements along the oblique lines. The curve CD is parabolic and D is the vertex of the curve. Hidden detail need not be shown.
Oxford Local Examinations (See Ch. 11 for information not in Ch. 6)

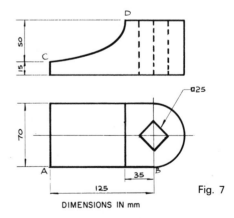

Fig. 7

DIMENSIONS IN mm

8. A special link for a mechanism has dimensions as shown in Fig. 8. Draw an oblique view of this link resting on the flat face, using an angle of 30°, with the centre line marked AB sloping upward to the right and with all dimensions *full size*. Radius curves may be sketched in and hidden details are to be omitted.

Oxford Local Examinations

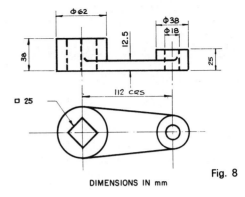

DIMENSIONS IN mm

Fig. 8

7
Enlarging and reducing plane figures and equivalent areas

Definition
Similar figures are figures that have the same shape but may be different in size.

Constructions
To construct a figure, similar to another figure, having sides 7/5 the length of the given figure.

Three examples, using the same basic method, are shown in Fig. 7/1.

Select a point P, sometimes called the centre of similitude, in one of the positions shown.

From P draw lines through all the corners of the figure.

Extend the length of one of the lines from P to a corner, say PQ, in the ratio 7:5. The new length is PR.

Beginning at R, draw the sides of the larger figure parallel to the sides of the original smaller figure.

This construction works equally well for reducing the size of a plane figure. Fig. 7/2 shows an irregular hexagon reduced to 4/9 its original size.

These constructions are practical only if the figure which has to be enlarged or reduced has straight sides. If the outline is irregular, a different approach is needed. Fig. 7/3 shows the face of a clown in two sizes, one twice that of the other. The change in size is determined by the two grids. A grid of known size is drawn over the first face and then another grid, similar to the first and at the required scale, is drawn alongside. Both grids are marked off, from A to J and from 1 to 5 in this

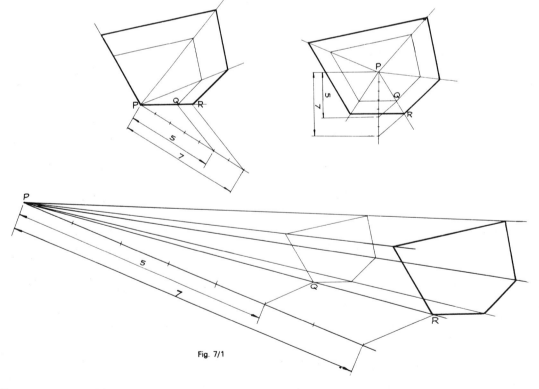

Fig. 7/1

case, and the points where the irregular outlines cross the lines of the grid are transferred from one grid to the other.

The closer together the lines of the grid, the greater the accuracy of the scaled copy.

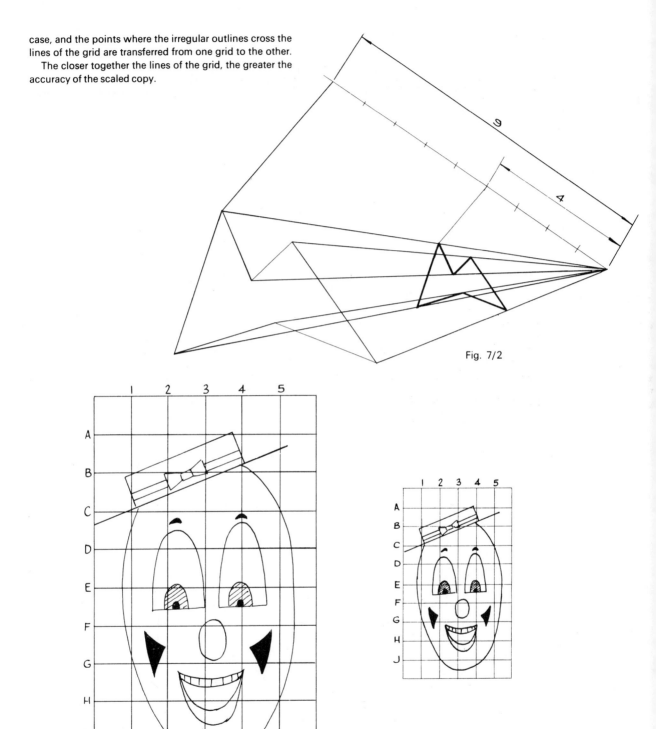

Fig. 7/2

Fig. 7/3

It is sometimes necessary to enlarge or reduce a plane figure in one direction only. In this case, although the dimensions are changed, the proportions remain the same. Fig. 7/4 shows a simple example of this. The figure has overall dimensions of 4 cm × 4 cm. The enlarged version retains the original proportions but now measures 6 cm × 4 cm.

First produce CA and BA. Mark off the new dimensions along CA and BA produced. This gives AB' and AC'.

Draw the square AB'XB and the rectangle ACYC' and draw the diagonals AX and AY.

From points along the periphery of the original plane figure (in this case 1 to 10), draw lines horizontally and vertically to and from the diagonals to intersect in 1', 2', 3', etc. Points 1' to 10' give the new profile.

Fig. 7/5 shows how a figure can be reduced on one side and enlarged on the other. A basic 50 mm × 50 mm shape has been changed proportionally into an 80 mm × 30 mm figure. Although this figure is more complicated than Fig. 7/4, with a corresponding increase in the number of points plotted, the basic construction is the same.

There is very little practical application of this type of construction these days. When plasterers produced flamboyant ceilings with complicated cornices, and carpenters had to make complex architraves and mouldings,

this type of construction was often employed. However, it is still a good exercise in plane geometry and does occasionally find an application.

The enlarged or reduced figures produced in Figs. 7/4 and 7/5 are mirror images of the original figures. Usually this does not matter, particularly if the figure is for a template; it just has to be turned over. However, if it does matter, a construction similar to that used in Fig. 7/6 must be used. In this case, a basic 60 mm × 40 mm shape has been changed into a 30 mm × 20 mm shape.

A'B' and A'C' are drawn parallel to AB and AC and marked off 20 mm and 30 mm long respectively.

AA' and BB', AA' and CC' are produced to meet in Q and P respectively.

The curved part of the figure is divided into as many parts as is necessary to produce an accurate copy.

The rest of the construction should be self-explanatory.

The transfer of the markings along A'B' and A'C' on the original figure to the required figure is made easier by the use of a 'trammel'; this is a rather pompous title for a piece of paper with a straight edge. If you lay this piece of paper along A'C' on the given figure and mark off A', C' and all the relevant points in between, you can line up the paper with A' and C' on the required figure and transfer the points between A' and C' onto the required figure. The same thing can be done for A'B'.

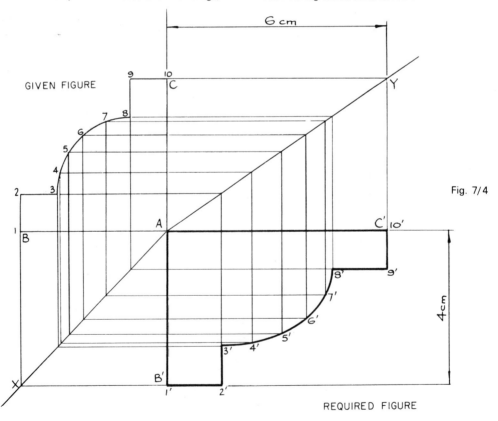

GIVEN FIGURE

Fig. 7/4

REQUIRED FIGURE

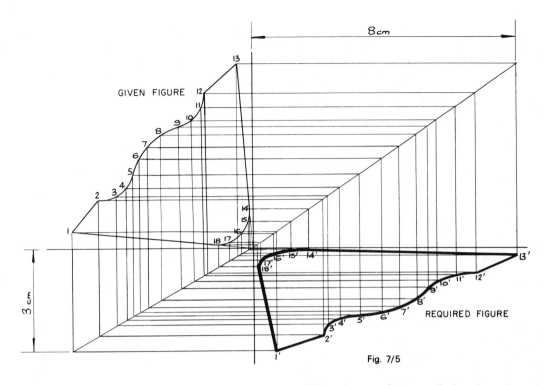

GIVEN FIGURE

8cm

3cm

REQUIRED FIGURE

Fig. 7/5

All the changes of shape so far have been dependent upon a known change of length of one or more of the sides. No consideration has been made of a specific change of area. The ability to enlarge or reduce a given shape in terms of area has applications. If, for instance, fluid flowing in a pipe is divided into two smaller pipes of equal area, then the area of the larger pipe will be twice that of the two smaller ones. This does not mean, of course, that the dimensions of the larger pipe are twice that of the smaller ones.

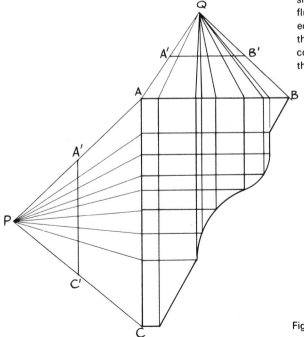

GIVEN FIGURE

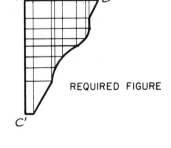

REQUIRED FIGURE

Fig. 7/6

Fig. 7/7 shows two similar constructions for enlarging a pentagon so that its new area is twice that of the original.

Select a point P. (This may be on a corner, or within the outline of the pentagon, or outside the outline although this is not shown because the construction is very large).

Let A be a corner of the given pentagon.

Join PA and produce it.

Draw a semi-circle, centre P, radius PA.

From P, drop a perpendicular to PA to meet the semi-circle in S.

Mark off PR : PQ in the required ratio, in this case 2 : 1.

Bisect AR in O, and erect a semi-circle, radius OR to cut PS produced in T.

Join SA and draw TA' parallel to SA.

A' is the first corner of the enlarged pentagon.

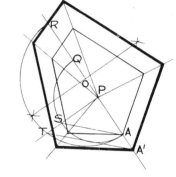

Fig. 7/7

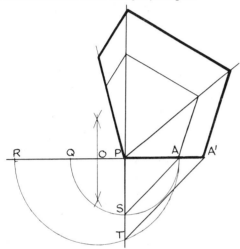

Although Fig. 7/7 shows a pentagon, the construction applies to any plane figure and can be used to increase and decrease a plane figure in a known ratio of areas. Fig. 7/8 shows a figure reduced to 4/7 its original size. The construction is identical to that used for Fig. 7/6 except that the ratio PR : PQ is 4 : 7. Note that if there is a circle or part circle in the outline, the position of its centre is plotted.

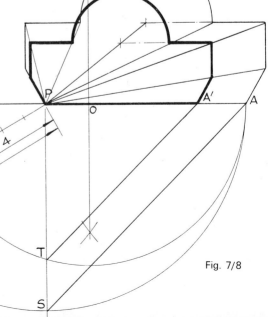

54

Fig. 7/8

EQUIVALENT AREAS

To construct a rectangle equal in area to a given triangle ABC (Fig. 7/9)

1. From B, the apex of the triangle, drop a perpendicular to meet the base in F.
2. Bisect FB.
3. From A and C erect perpendiculars to meet the bisected line in D and E.

ADEC is the required rectangle.

It should be obvious from the shading that the part of the triangle that is outside the rectangle is equal in area to that part of the rectangle that overlaps the triangle.

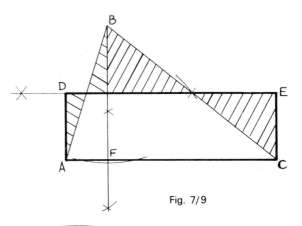

Fig. 7/9

To construct a square equal in area to a given rectangle. ABCD (Fig. 7/10)

1. With centre D, radius DC, draw an arc to meet AD produced in E.
2. Bisect AE and erect a semi-circle, radius AF, centre F.

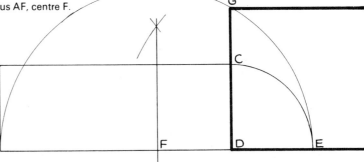

3. Produce DC to meet the semi-circle in G.

DG is one side of the square. (For the construction of a square, given one of the sides, see Ch. 2.)

This construction can be adapted to find the square root of a number. Fig. 7/11 shows how to find $\sqrt{6}$.

Since the area of the rectangle equals that of the square, then

$$ab = c^2$$

If a always $= 1$, then $b = c^2$

or $\sqrt{b} = c$

Thus, always draw the original rectangle with one side equal to one unit, and convert the rectangle into a square of equal area.

Fig. 7/10

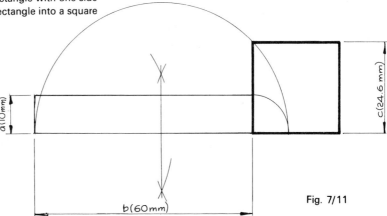

Fig. 7/11

To construct a square equal in area to a given triangle (Fig. 7/12)

This construction is a combination of those described in Figs. 7/9 and 7/10. First change the triangle into a rectangle of equivalent area and then change the rectangle into a square of equivalent area.

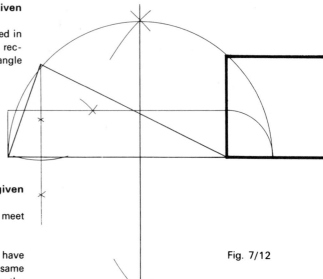

Fig. 7/12

To construct a triangle equal in area to a given polygon ABCDE (Fig. 7/13)

1. Join CE and from D draw a line parallel to CE to meet AE produced in F.
2. Join CF.

Since DF is parallel to CE, triangles CDE and CFE have the same base and vertical height and therefore the same area. The polygon ABCDE now has the same area as the quadrilateral ABCF and the original five-sided figure has been reduced to a four-sided figure of the same area.

3. Join CA and from B draw a line parallel to CA to meet EA produced in G.
4. Join CG.

Since BG is parallel to CA, triangles CBA and CGA have the same base and vertical height and therefore the same area. The quadrilateral ABCF now has the same area as the triangle GCF and the original five-sided figure has been reduced to a three-sided figure of the same area.

GCF is the required triangle.

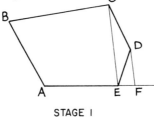

STAGE 1

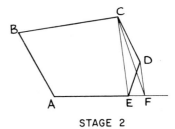

STAGE 2

Fig. 7/13

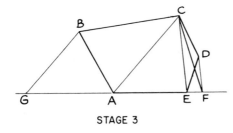

STAGE 3

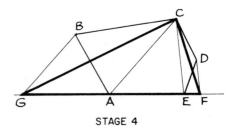

STAGE 4

The theorem of Pythagoras says that 'in a right-angled triangle, the square on the hypotenuse is equal to the sum of the squares on the other two sides'. When this theorem is shown pictorially, it is usually illustrated by a triangle with squares drawn on the sides. This tends to be a little misleading since the theorem is valid for **any** similar plane figures (see Fig. 7/14).

This construction is particularly useful when you wish to find the size of a circle which has the equivalent area of two or more smaller circles added together.

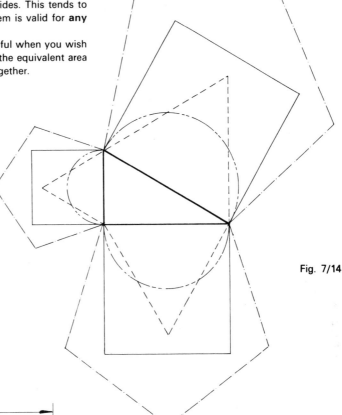

Fig. 7/14

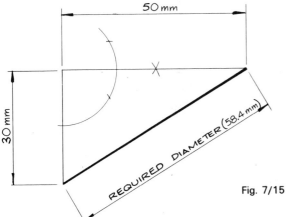

Fig. 7/15

To find the diameter of a circle which has the same area as two circles, 30 mm and 50 mm diameter (Fig. 7/15)

Draw a line 30 mm long.

From one end erect a perpendicular, 50 mm long.

The hypotenuse of the triangle thus formed is the required diameter (58.4 mm).

If you have to find the single equivalent diameter of more than two circles, reduce them in pairs until you have two, and then finally one left.

57

To divide a triangle ABC into three parts of equal area by drawing lines parallel to one of the sides (say BC) (Fig. 7/16)

1. Bisect AC (or AB) in O and erect a semi-circle, centre O, radius OA.
2. Divide AC into three equal parts AD, DE and EC and erect perpendiculars from D and E to meet the semi-circle in D₁ and E₁.
3. With centre A, and radius AD₁, draw an arc to cut AC in D₂.
4. With centre A, and radius AE₁, draw an arc to cut AC in E₂.
5. From D₂ and E₂ draw lines parallel to BC to meet AB in D₃ and E₃ respectively.

The areas AD_2D_3, $D_3D_2E_2E_3$, E_3E_2CB are equal.

Although Fig. 7/16 shows a triangle divided into three equal areas, the construction can be used for any number of equal areas.

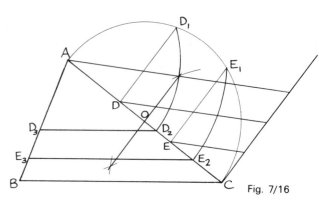

Fig. 7/16

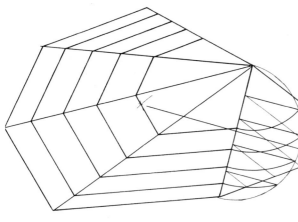

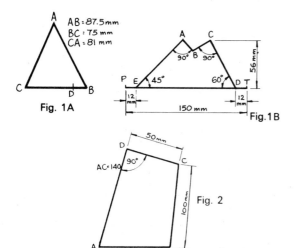

Fig. 7/17

To divide a polygon into a number of equal areas (say 5) by lines drawn parallel to the sides

This construction is very similar to that used for Fig. 7/16. Proceed as for the triangle and complete as shown in Fig. 7/17.

Again, this construction can be used for any polygon and can be adapted to divide any polygon into any number of equal areas.

Exercises 7

(All questions originally set in Imperial units)

1. (a) Construct the triangle ABC, shown in Fig. 1A, from the information given and then construct a triangle CDA in which CD and CB are in the ratio 5 : 7.

   (b) Construct the polygon ABCDE shown in Fig. 1B. Construct by the use of radiating lines a polygon PQRST similar to ABCDE so that both polygons stand on the line PT.
   Southern Regional Examinations Board

2. Fig. 2 shows a sail for a model boat. Draw the figure, full size, and construct a similar shape with the side corresponding to AB 67 mm long.
   Middlesex Regional Examining Board

3. Without the use of a protractor or set squares construct a polygon ABCDE standing on a base AB given that AB = 95 mm, BC = 75 mm, CD = 55 mm, AE = 67.5 mm, ∠ABC = 120°, ∠EAB = 82½°, and ∠CDE = 90°. Also construct a similar but larger polygon so that the side corresponding with AB becomes 117.5 mm. Measure and state the lengths of the sides of the enlarged polygon.
   Oxford Local Examinations

AB = 87.5 mm
BC = 75 mm
CA = 81 mm

Fig. 1A

90° B 90°
P E 45°   60° D T
12 mm   150 mm   12 mm   56 mm

Fig. 1B

50 mm
D 90°
AC = 140
C
100 mm
A 88 mm B

Fig. 2

4. Make a copy of the plane figure shown in Fig. 3. Enlarge your figure proportionally so that the base AB measures 88 mm.
University of London School Examinations

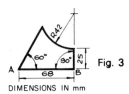
Fig. 3
DIMENSIONS IN mm

5. In the triangle ABC, AB = 82 mm, BC = 105 mm, and CA = 68 mm. Draw a triangle similar to ABC, and having an area one-fifth the area of ABC.
Oxford and Cambridge Schools Examinations Board

6. Draw a polygon ABCDEF making AB = 32 mm, BC = 38 mm, CD = 50 mm, DE = 34 mm, EF = 28 mm, FA = 28 mm, AC = 56 mm, AD = 68 mm, and AE = 50 mm.
Construct a further polygon similar to ABCDEF but having an area larger in the ratio of 4 : 3.
Cambridge Local Examinations

7. Construct, full size, the figure illustrated in Fig. 4 and by radial projection, superimpose about the same centre a similar figure whose area is three times as great as that shown in Fig. 4.
Oxford Local Examinations

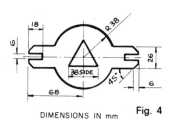
DIMENSIONS IN mm
Fig. 4

8. Fig. 5 shows a section through a length of moulding. Draw an enlarged section so that the 118 mm dimensions becomes 172 mm.
Oxford Local Examinations (See Ch. 11 for information not in Ch. 7)

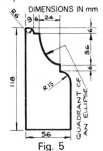
DIMENSIONS IN mm
Fig. 5

9. Fig. 6 shows a shaped plate, of which DE is a quarter of an ellipse. Draw:
(a) the given view, full size;
(b) an enlarged view of the plate so that AB becomes 150 mm and AC 100 mm. The distances parallel to AB and AC are to be enlarged proportionately to the increase in length of AB and AC respectively.
Oxford Local Examinations (See Ch. 11 for information not in Ch. 7)

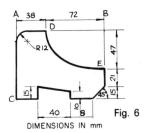
DIMENSIONS IN mm
Fig. 6

10. The shape is shown in Fig. 7 of a plate of uniform thickness.
Draw the figure and, with one corner in the position shown, add a square which would represent the position of a square hole reducing the weight of the plate by 25%.
Oxford and Cambridge Schools Examinations Board

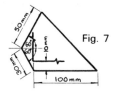
Fig. 7

11. Construct a regular hexagon having a distance between opposite sides of 100 mm. Reduce this hexagon to a square of equal area. Measure and state the length of side of this square.
Joint Matriculation Board

12. A water main is supplied by two pipes of 75 mm and 100 mm diameter. It is required to replace the two pipes with one pipe which is large enough to carry the same volume of water.
Part 1. Draw the two pipes and then, using a geometrical construction, draw the third pipe.
Part 2. Draw a pipe equal in area to the *sum* of the three pipes.
Southern Regional Examinations Board

13. Three squares have side lengths of 25 mm, 37.5 mm and 50 mm respectively. Construct, without resorting to calculations, a single square equal in area to the three squares, and measure and state the length of its side.
Cambridge Local Examinations

8

The blending of lines and curves

It is usually only the very simple type of engineering detail that has an outline composed entirely of straight lines. The inclusion of curves within the outline of a component may be for several reasons: to eliminate sharp edges, thereby making it safer to handle; to eliminate a stress centre, thereby making it stronger; to avoid extra machining, thereby making it cheaper; and last, but by no means least, to improve its appearance. This last reason applies particularly to those industries which manufacture articles to sell to the general public. It is not enough these days to make vacuum cleaners, food mixers or ball point pens functional and reliable. It is equally important that they be attractive so that they, and not the competitors' products are the ones that catch the shopper's eye. The designer uses circles and curves to smooth out and soften an outline. Modern machine shop processes like cold metal forming, and the increasing use of plastics and laminates, allow complex outlines to be manufactured as cheaply as simple ones, and the blending of lines and curves plays an increasingly important role in the draughtsman's world.

Blending is a topic that students often have difficulty in understanding and yet there are only a few ways in which lines and curves can be blended. When constructing an outline which contains curves blending, do not worry about the point of contact of the curves; rather be concerned with the positions of the centres of the curves. A curve will not blend properly with another curve or line unless the centre of the curve is correctly found. If the centre is found exactly, the curve is bound to blend exactly.

To find the centre of an arc, radius _r_, which blends with two straight lines meeting at right angles (Fig. 8/1)
With centre A, radius _r_, draw arcs to cut the lines of the angle in B and C.

With centres B and C, radius _r_, draw two arcs to intersect in 0.
0 is the required centre.

This construction applies only if the angle is a right angle. If the lines meet at any angle other than 90°, use the construction shown in Fig. 8/2.

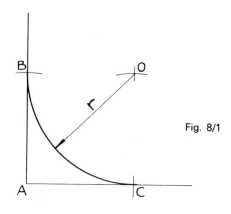

Fig. 8/1

To find the centre of an arc, radius *r*, which blends with two straight lines meeting at any angle (Fig. 8/2)
Construct lines, parallel with the lines of the angle and distance *r* away, to intersect in O.
O is the required centre.

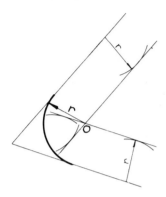
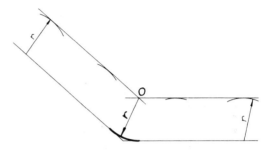

Fig. 8/2

To find the centre of an arc, radius *r*, which passes through a point P and blends with a straight line (Fig. 8/3)
Construct a line, parallel with the given line, distance *r* away. The centre must lie somewhere along this line.
    With centre P, radius *r*, draw an arc to cut the parallel line in O.
O is the required centre.

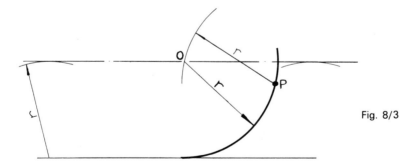

Fig. 8/3

To find the centre of an arc, radius _R_, which blends with a line and a circle, centre B, radius _r_

There are two possible centres, shown in Figs. 8/4 and 8/5.

Construct a line, parallel with the given line, distance _R_ away. The centre must lie somewhere along this line.

With centre B, radius _R_ + _r_, draw an arc to intersect the parallel line in O.

O is the required centre.

The alternative construction is:

Construct a line, parallel with the given line, distance _R_ away. The centre must lie somewhere along this line.

With centre B, radius _R_ − _r_, draw an arc to intersect the parallel line in O.

O is the required centre.

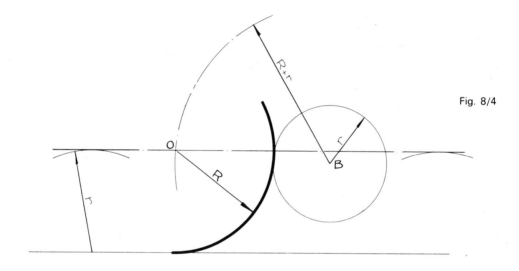

Fig. 8/4

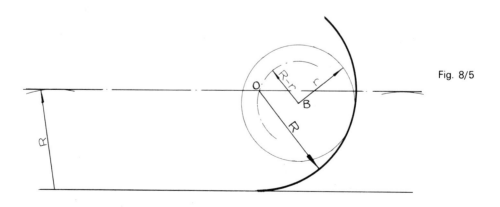

Fig. 8/5

To find the centre of an arc, radius R, which blends with two circles, centres A and B, radii r_1 and r_2 respectively

There are two possible centres, shown in Figs. 8/6 and 8/7.

If an arc, radius R, is to blend with a circle, radius r, the centre of the arc must be distance R from the circumference and hence $R + r$ (Fig. 8/6) or $R - r$ (Fig. 8/7) from the centre of the circle.

With centre A, radius $R + r_1$, draw an arc.

With centre B, radius $R + r_2$, draw an arc to intersect the first arc in O.

O is the required centre.

The alternative construction is:

With centre A, radius $R - r_1$, draw an arc.

With centre B, radius $R - r_2$, draw an arc to intersect the first arc in O.

O is the required centre.

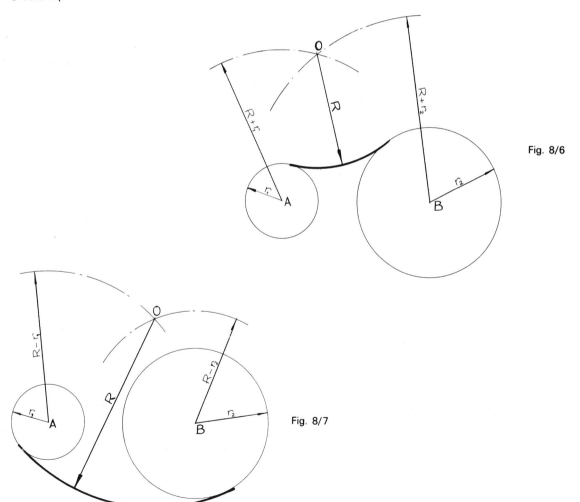

Fig. 8/6

Fig. 8/7

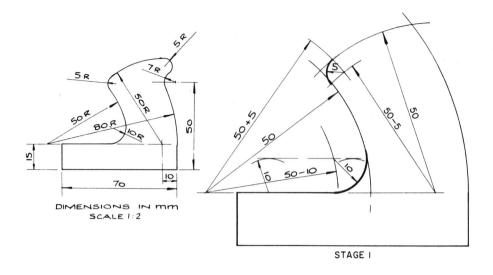

DIMENSIONS IN mm
SCALE 1:2

STAGE I

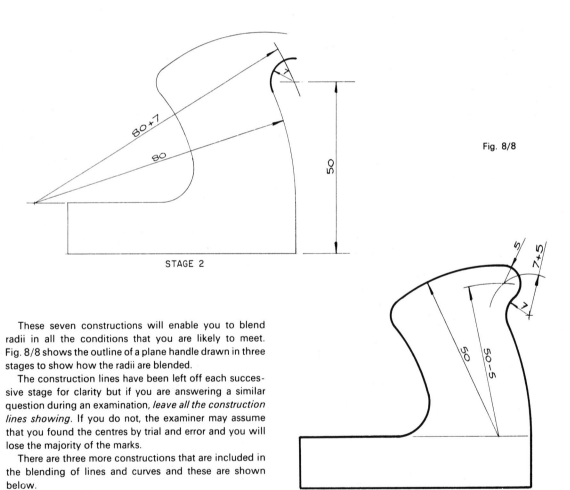

STAGE 2

Fig. 8/8

STAGE 3

These seven constructions will enable you to blend radii in all the conditions that you are likely to meet. Fig. 8/8 shows the outline of a plane handle drawn in three stages to show how the radii are blended.

The construction lines have been left off each successive stage for clarity but if you are answering a similar question during an examination, *leave all the construction lines showing*. If you do not, the examiner may assume that you found the centres by trial and error and you will lose the majority of the marks.

There are three more constructions that are included in the blending of lines and curves and these are shown below.

To join two parallel lines with two equal radii, the sum of which equals the distance between the lines (Fig. 8/9)

Draw the centre line between the parallel lines.

From a point A, drop a perpendicular to meet the centre line in O_1.

With centre O_1, radius O_1A, draw an arc to meet the centre line in B.

Produce AB to meet the other parallel line in C.

From C erect a perpendicular to meet the centre line in O_2.

With centre O_2, radius O_2C, draw the arc BC.

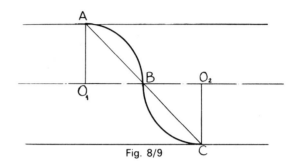

Fig. 8/9

To join two parallel lines with two equal radii, r, the sum of which is greater than the distance between the lines (Fig. 8/10)

Draw the centre line between the parallel lines.

From a point A, drop a perpendicular and on it mark off $AO_1 = r$.

Draw the centre line between the parallel lines.

From a point A, drop a perpendicular and on it mark off $AO_1 = r$.

With centre O_1, radius r, draw an arc to meet the centre line in B.

Produce AB to meet the other parallel line in C.

From C erect a perpendicular $CO_2 = r$.

With centre O_2, radius r, draw the arc BC.

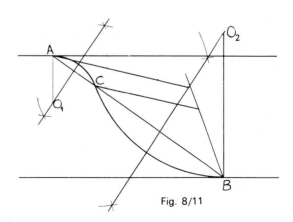

Fig. 8/10

To join two parallel lines with two unequal radii (say in the ratio of 3 : 1) given the ends of the curve A and B (Fig. 8/11)

Join AB and divide into the required ratio, AC : CB = 1 : 3.

Perpendicularly bisect AC to meet the perpendicular from A in O_1.

With centre O_1, radius O_1A, draw the arc AC.

Perpendicularly bisect CB to meet the perpendicular from B in O_2.

With centre O_2 and radius O_2B, draw the arc CB.

Fig. 8/11

Exercises 8

(All questions originally set in Imperial units)

1. Fig. 1 shows an exhaust pipe gasket. Draw the given view full size and show any constructions used in making your drawing. Do not dimension your drawing.
Southern Regional Examinations Board

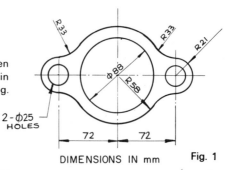

2-Φ25
HOLES

72   72

DIMENSIONS IN mm   **Fig. 1**

2. Fig. 2 is an elevation of the turning handle of a can opener. Draw this view, twice full size, showing clearly the method of establishing the centres of the arcs.
East Anglian Examinations Board

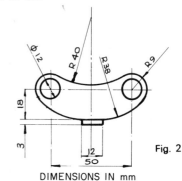

DIMENSIONS IN mm   **Fig. 2**

3. Fig. 3 shows the outline of an electric lamp.
Important—Construction lines must be visible, showing clearly how you obtained the centres of the arcs and the exact positions of the junctions between arcs and straight lines.
Part 1. Draw the shape *full size*.
Part 2. Line A–B is to be increased to 28 mm. Construct a scale and using this scale draw the left half or right half of the shape, increasing all other dimensions proportionally.
Southern Regional Examinations Board

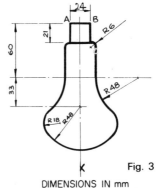

Fig. 3

DIMENSIONS IN mm

4. Fig. 4 shows a garden hoe. Draw this given view *full size* and show any construction lines used in making the drawing. Do not dimension the drawing.
Southern Regional Examinations Board

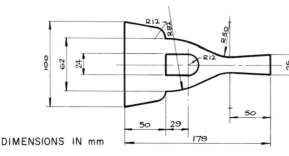

DIMENSIONS IN mm

Fig. 4

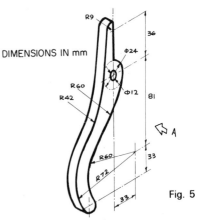

DIMENSIONS IN mm

5. Fig. 5 shows one half of a pair of pliers. Draw, *full size*, a front elevation looking from A. Your constructions for finding the centres of the arcs must be shown.
South-East Regional Examinations Board

Fig. 5

6. Fig. 6 shows the design for the profile of a sea wall. Draw the profile of the sea wall to a scale of 10 mm = 2m. Measure in metres the dimensions A, B, C and D and insert these on your drawing. In order to do this you should construct an open divided scale of 10 mm = 2 m to show units of 1 m.
Constructions for obtaining the centres of the radii must be clearly shown.
Metropolitan Regional Examinations Board

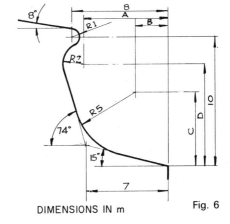

DIMENSIONS IN m   Fig. 6

7. Details of a spanner for a hexagonal nut are shown in Fig. 7. Draw this outline showing clearly all constructions. Scale: full size.
Oxford Local Examinations

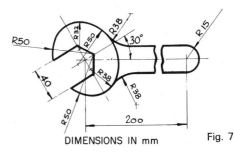

DIMENSIONS IN mm   Fig. 7

8. The end of the lever for a safety valve is shown in Fig. 8. Draw this view, showing clearly all construction lines. Scale: ½ full size.
Oxford Local Examinations

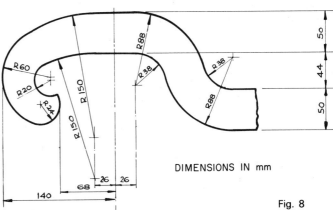

DIMENSIONS IN mm

Fig. 8

9. Draw, to a scale of 2 : 1, the front elevation of a rocker arm as illustrated in Fig. 9.
Oxford Local Examinations

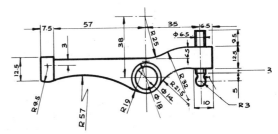

DIMENSIONS IN mm   Fig. 9

67

9
Loci

Definition

A locus (plural *loci*) is the path traced out by a point which moves under given definite conditions.

You may not have been aware of it, but you have met loci many times before. One of the most common loci is that of a point which moves so that its distance from another fixed point remains constant: this produces a circle. Another locus that you know is that of a point which moves so that its distance from a line remains constant: this produces parallel lines.

Problems on loci can take several different forms. One important practical application is finding the path traced out by points on mechanisms. This may be done simply to see if there is sufficient clearance around a mechanism or, with further knowledge beyond the scope of this book, to determine the velocity and hence the forces acting upon a component.

There are very few rules to learn about loci; it is mainly a subject for common sense. A locus is formed by continuous movement and you have to 'stop' the movement several times and find and plot the position of the point that you are interested in. Take, for instance, the case of the man who was too lazy to put wedges under his ladder. The inevitable happened and the ladder slipped. The path that the feet of the man took is shown in Fig. 9/1.

The top of the ladder slips from T to T_9. The motion of the top of the ladder has been stopped at T_1, T_2, T_3, etc. and, since the length of the ladder remains constant, the corresponding positions of the bottom of the ladder, B_1, B_2, B_3, etc. can be found. The positions of the ladder T_1B_1, T_2B_2, T_3B_3, etc. are drawn and the position of the man's feet, 1, 2, 3, etc., are marked. The points are joined together with a smooth curve. It is interesting to note that the man hits the ground at right angles (assuming that he remains on the ladder). The resulting jar often causes serious injury and is one of the reasons for using chocks.

Another simple example is the locus of the end of a

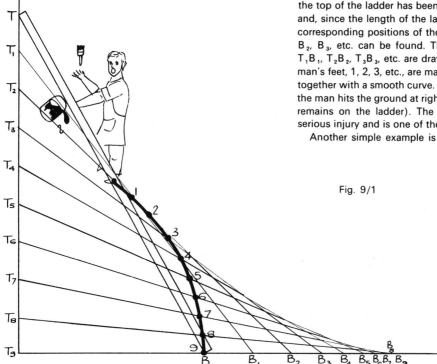

Fig. 9/1

68

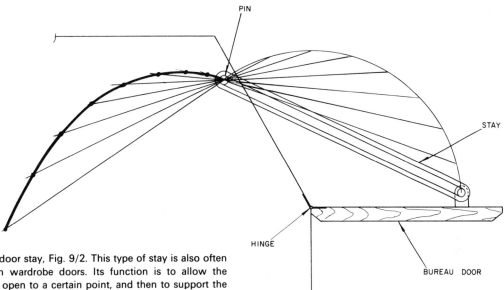

PIN

STAY

HINGE

BUREAU DOOR

Fig. 9/2

bureau door stay, Fig. 9/2. This type of stay is also often used on wardrobe doors. Its function is to allow the door to open to a certain point, and then to support the door in that position.

The stay, of course, has two ends and the locus on one end is easily found: it is an arc whose centre is the hinge. The other end of the stay is allowed to slide through the pin but it is not allowed to move off it. As the end of the stay moves along the arc, its movement is stopped several times and the position of the other end of the stay is marked. These points are joined together with a smooth curve. Obviously the designer of such a bureau would have to plot this locus before deciding the depth of the bureau.

Loci of mechanisms
The bureau door stay is a very simple mechanism. We now look at some of the loci that can be found on the moving parts of some machines.

Definitions
Velocity is speed in a given direction. It is a term usually reserved for inanimate objects; we talk about the muzzle velocity of a rifle or the escape velocity of a space probe. When we use the word speed we refer only to the rate of motion. When we use the word velocity we refer to the rate of motion and the direction of the motion.

Linear velocity is velocity along a straight line (a linear graph is a straight line).

Angular velocity is movement through a certain angle in a certain time. It makes no allowance for distance travelled. If, as in Fig. 9/3, a point P moves through 60° in 1 second, its angular velocity is exactly the same as that of Q, providing that Q also travels through 60° in 1 second. The velocity, as distinct from the angular velocity, will be much greater of course.

Constant velocity, linear or angular, is movement without acceleration or deceleration.

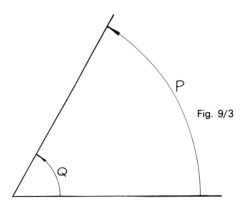

P

Q

Fig. 9/3

The piston/connecting rod/crank mechanism is very widely used, principally because of its application in internal combustion engines. The piston travels in a straight line; the crank rotates. The connecting rod, which links these two, follows a path which is somewhere in between these two, the exact shape being dependent on the point of the rod being considered. Fig. 9/4 shows

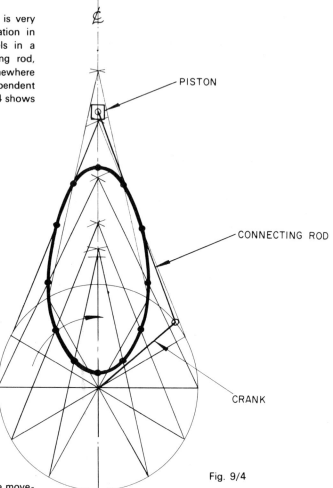

PISTON

CONNECTING ROD

CRANK

Fig. 9/4

the locus of a point half way up (or down) the rod.

The movement of the crank is continuous. The movement of the piston is also continuous between the top and bottom of its travel. This movement, as before, must be 'stopped' several times and the positions of the centre of the connecting rod found. As with most machines that have cranks, the best policy is to plot the position of the crank in twelve equally spaced positions. This is easily achieved with a 60° set square. The piston must always lie on the centre line and, of course, the connecting rod cannot change its length. It is therefore a simple matter to plot the position of the connecting rod for the twelve positions of the crank. This is best done with compasses or dividers. The mid-point of the connecting rod can then be marked with dividers and the points joined together with a smooth curve.

The direction of rotation of the crank is usually given in problems of this nature. It may make no difference to plotting any of the loci but it could make a tremendous difference to the functioning of the real machine: what good is a car that does 120 km/h backwards and 10 km/h forwards?

70

Trammels

A trammel can consist of a piece of paper or a piece of
card or even the edge of a set square. It must have a
straight edge and you must be able to mark it with a pencil.
A trammel enables you to plot a locus more quickly than
the method shown above. However, if you are intending
to sit the G.C.E. or C.S.E. examination check that the
syllabus allows you to use trammels.

Fig. 9/5 shows a crank/rod/slider mechanism where
a point $\frac{1}{3}$ along the rod has been plotted for one revolution
of the crank. The length of the rod, and the point, are
marked on a piece of paper. One end of the rod is con-
strained to travel around the crank circle and the other
slides up and down the centre line of the slider. Move the
trammel so that one end is always on the circle whilst
the other end is always on the slider centre line, marking
the required point as many times as necessary. Join the
points with a smooth curve.

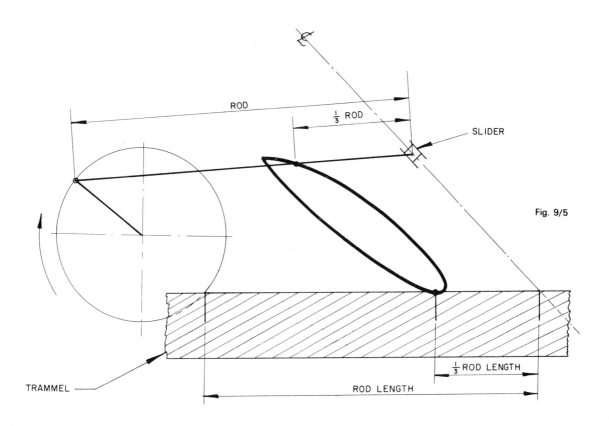

Fig. 9/5

On mechanisms that are more complicated, it is some-times necessary to plot one locus to obtain another. Fig. 9/6 shows a mechanism which consists of two cranks, O_1A and O_2B, and two links, AB and CD. It is required to plot the locus of P, a point on the lower end of the link CD.

Before we can plot any positions of the link CD, and hence P, we must know the position of C at any given moment. This can only be done by plotting the locus of C, ignoring the link CD. Once this has been done, we can find the position of CD at any given moment, and hence the locus of point P.

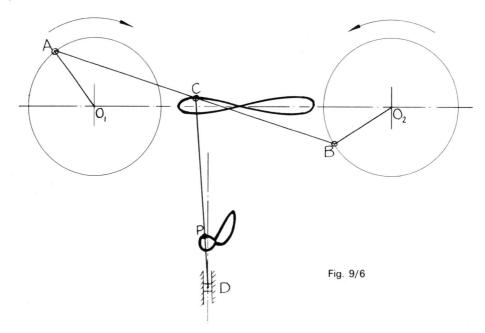

Fig. 9/6

No construction lines are shown in Fig. 9/6 since they would only make the drawing even more confusing! The locus of P could have been found by 'stopping' one of the cranks in twelve different places and finding the twelve new positions of the link AB. This, in turn, would enable you to find twelve positions for C, and then twelve positions of the rod CD. Finally this would lead to the twelve required positions of P. Alternatively, the locus of C could have been plotted with a trammel which had the length of the link AB, and the position of C marked on it. Another trammel, with the length of the rod CD and the position of P marked on it, would have given the locus of P.

Some Other Problems in Loci

A locus is defined as the path traced out by a point which moves under given definite conditions. Three examples of loci are shown below where a point moves relative to another point or to lines.

To plot the locus of a point P which moves so that its distance from a point S and a line X–Y is always the same (Fig. 9/7)

The point S is 20 mm from the line X–Y.

The first point to plot is the one that lies between S and the line. Since S is 20 mm from the line, and P is equidistant from both, this first point is 10 mm from both.

If we now let the point P be 20 mm from S, it will lie somewhere on the circumference of a circle, centre S, radius 20 mm. Since the point is equidistant from the line, it must also lie on a line drawn parallel to X–Y and 20 mm away. The second point, then, is the intersection of the 20 mm radius arc and the parallel line.

The third point is at the intersection of an arc, radius 30 mm and centre S, and a line drawn parallel to X–Y and 30 mm away.

The fourth point is 40 mm from both the line and the point S. This may be continued for as long as is required.

The curve produced is a parabola.

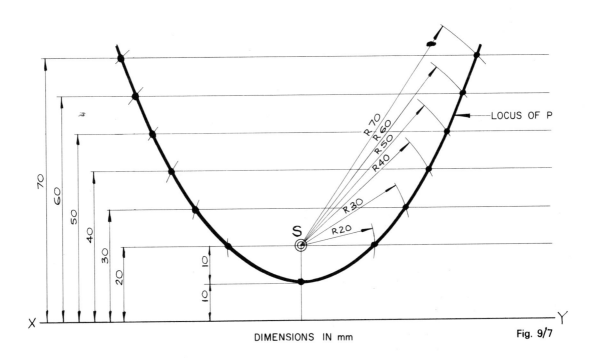

DIMENSIONS IN mm

Fig. 9/7

To plot the locus of a point P which moves so that its distance from two fixed points R and S, 50 mm apart, is always in the ratio 2:1 respectively (Fig. 9/8)

As in the previous example, the first point to plot is the one that lies between R and S. Since it is twice as far away from R as it is from S, this is done by proportional division of the line RS.

If we now let P be 40 mm from R it must be 20 mm from S. Thus, the second position of P is at the intersection of an arc, centre R, radius 40 mm and another arc, centre S and radius 20 mm.

The third position of P is the intersection of arcs, radii 50 mm and 25 mm, centres R and S respectively.

This is continued for as long as necessary. In this case, at a point 100 mm from R and 50 mm from S, the locus meets itself to form a circle.

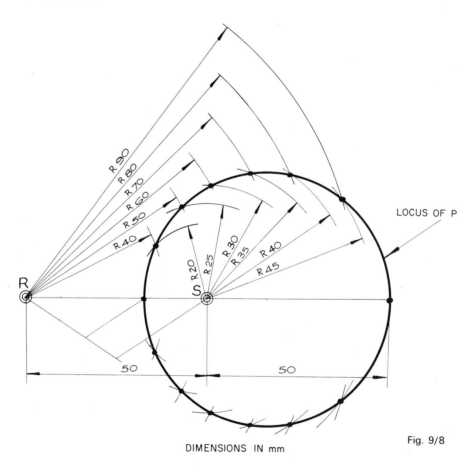

DIMENSIONS IN mm

Fig. 9/8

To plot the locus of a point P which moves so that its distance from the circumference of two circles, centres O_1 and O_2 and radii 20 mm and 15 mm respectively, is always in the ratio 2:3 respectively (Fig. 9/9)

As with the two previous examples, the first point to plot is the one that lies between the two circles. Thus, divide the space between the two circumferences in the ratio 3:2 by proportional division.

If we now let P be 10 mm from the circumference of the circle, Centre O_1, it will lie somewhere on a circle, centre O_1, and radius 30 mm. If it is 10 mm from the circumference of the circle, centre O_1, it will be 15 mm from the circumference of the circle, centre O_2, since the ratio of the distances of P from the circumferences of the circles is 3:2. Thus, the second position of P is the intersection of two arcs, radii 30 mm and 30 mm, centres O_1 and O_2.

The third position of P is the intersection of two arcs, radii 35 mm and 37.5 mm, centres O_1 and O_2 respectively.

The fourth position is at the intersection of arcs of 40 mm and 45 mm radii.

This is continued for as long as is required.

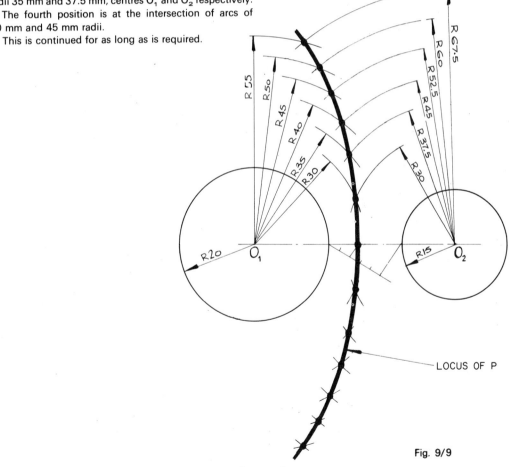

LOCUS OF P

Fig. 9/9

DIMENSIONS IN mm

Exercises 9

(All questions originally set in Imperial units)

1. Fig. 1 shows a door stay as used on a wardrobe door. The door is shown in the fully open position. Draw, *full size*, the locus of end A of the stay as the door closes to the fully closed position. The stay need only be shown diagrammatically as in Fig. 1A.

West Midlands Examinations Board

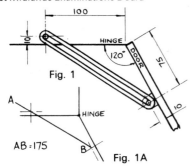

Fig. 1

AB = 175

Fig. 1A

DIMENSIONS IN mm

2. Fig. 2 shows a sketch of the working parts, and the working parts represented by lines, of a moped engine. Using the line diagram only, and drawing in single lines only, plot, full size, the locus of the point P for one full turn of the crank BC.

Do not attempt to draw the detail shown in the sketch.

Show all construction.

The trammel method must not be used.

East Anglian Examinations Board

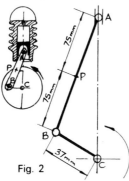

Fig. 2

3. In Fig. 3 the crank C rotates in a clockwise direction. The rod PB is connected to the crank at B and slides through the pivot D.

Plot, to a scale 1½ *full size*, the locus of P for one revolution of the crank.

South-East Regional Examinations Board

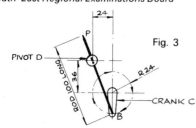

Fig. 3

DIMENSIONS IN mm

76

4. In Fig. 4 the stay BHA is pin-pointed at H and is free to rotate about the fixed point B. Plot the locus of P as end A moves from A to A'.

North Western Secondary School Examinations Board

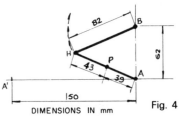

DIMENSIONS IN mm

Fig. 4

5. In Fig. 5, rollers 1 and 2 are attached to the angled rod. Roller 1 slides along slot AB while roller 2 slides along CD and back.

Draw, *full size*, the locus of P, the end of the rod, for the complete movement of roller 1 from A to B.

South-East Regional Examinations Board

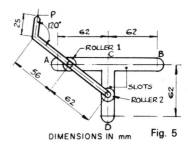

DIMENSIONS IN mm

Fig. 5

6. As an experiment a very low gear has been fitted to a bicycle. This gear allows the bicycle to move forward 50 mm for every 15 degrees rotation of the crank and pedal. These details are shown in Fig. 6.

(a) Draw, half full size, the crank and pedal in position as it rotates for every 50 mm forward motion of the bicycle up to a distance of 600 mm. The first forward position has been shown on the drawing.

(b) Draw a smooth freehand curve through the positions of the pedal which you have plotted.

(c) From your drawing find the angle of the crank OA to the horizontal when the bicycle has moved forward 255 mm.

Metropolitan Regional Examinations Board

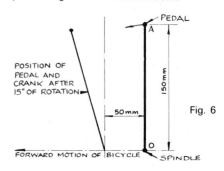

Fig. 6

7. Fig. 7 is a line diagram of a slotted link and crank of a shaping machine mechanism. The link AC is attached to a fixed point A about which it is free to move about the fixed point on the disc. The disc rotates about centre O. Attached to C and free to move easily about the points C and D is the link CD. D is also free to slide along DE.

When the disc rotates in the direction of the arrow, plot the locus of C, the locus of P on the link CD, and clearly show the full travel of B on AC.

Southern Universities' Joint Board

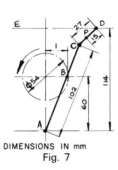

DIMENSIONS IN mm
Fig. 7

8. In Fig. 8, MP and NP are rods hinged at P, and A and D are guides through which MP and NP are allowed to move. D is allowed to move along BC, but rod NP is always perpendicular to BC. The guide A is allowed to rotate about its fixed point. Draw the locus of P above AB for all positions and when P is always equidistant from A and BC.

This locus is part of a recognised curve. Name the curve and the parts used in its construction.

Southern Universities' Joint Board

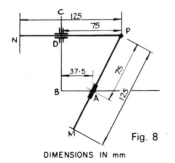

DIMENSIONS IN mm
Fig. 8

9. In the mechanism shown in Fig. 9, OA rotates about O, PC is pivoted at P, and QB is pivoted at Q. BCDE is a rigid link. OA = PC = CD = DE = 25 mm, BC = 37.5 mm, QB = 50 mm and AD = 75 mm. Plot the complete locus of E.

Oxford and Cambridge Schools Examinations Board

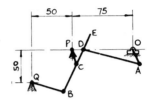

DIMENSIONS IN mm
Fig. 9

10. A rod AB 70 mm long rotates at a uniform rate about end A. Plot the path of a slider S, initially coincident with A, which slides along the rod, at a uniform rate, from A to B and back to A during one complete revolution of the rod.

Joint Matriculation Board

11. With a permanent base of 100 mm, draw the locus of the vertices of all the triangles with a constant perimeter of 225 mm.

Oxford and Cambridge Schools Examination Board

12. Three circles lie in a plane in the positions shown in Fig. 10. Draw the given figure and plot the locus of a point which moves so that it is always equidistant from the circumferences of circles A and B. Plot also the locus of a point which moves in like manner between circles A and C.

Finally draw a circle whose circumference touches the circles A, B and C and measure and state its diameter.

Cambridge Local Examinations Board

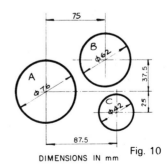

DIMENSIONS IN mm
Fig. 10

10
Orthographic projection

Orthographic projection is the solution to the biggest problem that a draughtsman has to solve—how to draw, with sufficient clarity, a three-dimensional object on a two-dimensional piece of paper. The drawing must show quite clearly the detailed outlines of all the faces and these outlines must be fully dimensioned. If the object is very simple, this may be achieved with a freehand sketch. A less simple object could be drawn in either isometric or oblique projections, although both these systems have their disadvantages. Circles and curves are difficult to draw in either system and neither shows more than three sides of an object in any one view. Orthographic projection, because of its flexibility in allowing any number of views of the same object, has none of these drawbacks.

Orthographic projection has two forms: First angle and Third angle; we shall discuss both. Traditionally, British industry has used 1st angle whilst the United States of America and, more recently, the Continental countries used the 3rd angle system. There is no doubt that British industry is rapidly changing to the 3rd angle system and, whilst this will take some years to complete, 3rd angle will eventually be the national and international standard of orthographic projection.

Fig. 10/1 shows a stepped block suspended between two planes. A plane is a perfectly flat surface. In this case one of the planes is horizontal and the other is vertical. The view looking on the top of the block is drawn directly above the block on the horizontal plane. The view looking on the side of the block is drawn directly in line with the block on the vertical plane. If you now take away the stepped block and, imagining that the two planes are hinged, fold back the horizontal plane so that it lines up with the vertical plane, you are left with two drawings of the block. One is a view looking on the top of the block and this is directly above another view looking on the side of the block. These two views are called elevations.

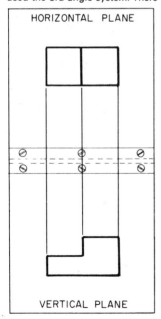

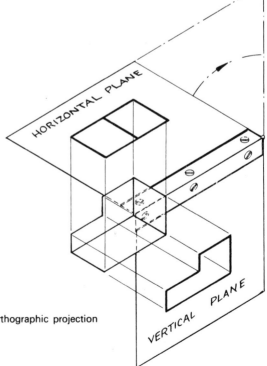

Fig. 10/1 3rd angle orthographic projection

Fig. 10/1 shows the block in 3rd angle orthographic projection. The same block is drawn in Fig. 10/2 in 1st angle orthographic projection. You still have a vertical and a horizontal plane but they are arranged differently. The block is suspended between the two planes and the view of the top of the block is drawn on the horizontal plane and the view of the side is drawn on the vertical plane. Again, imagining that the planes are hinged, the horizontal plane is folded down so that the planes are in line. This results in the drawing of the side of the block being directly above the drawing of the top of the block (compare this with the 3rd angle drawings).

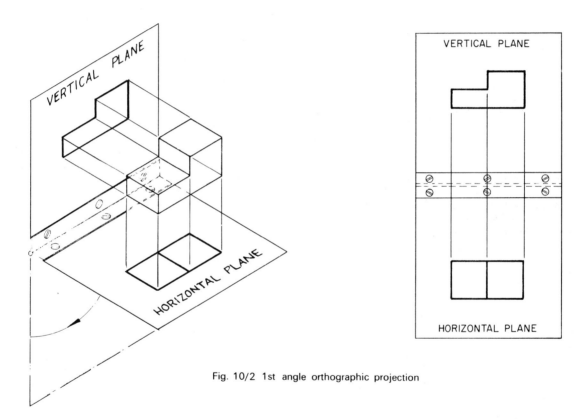

Fig. 10/2 1st angle orthographic projection

79

The reason why these two systems are called 1st and 3rd angle is shown in Fig. 10/3. If the horizontal plane (H.P.) and the vertical plane (V.P.) intersect as shown, it produces four quadrants. The first quadrant, or first angle, is the top right and the third is the bottom left. If the block is suspended between the V.P. and the H.P. in the first and third angles you can see how the views are projected onto the two planes.

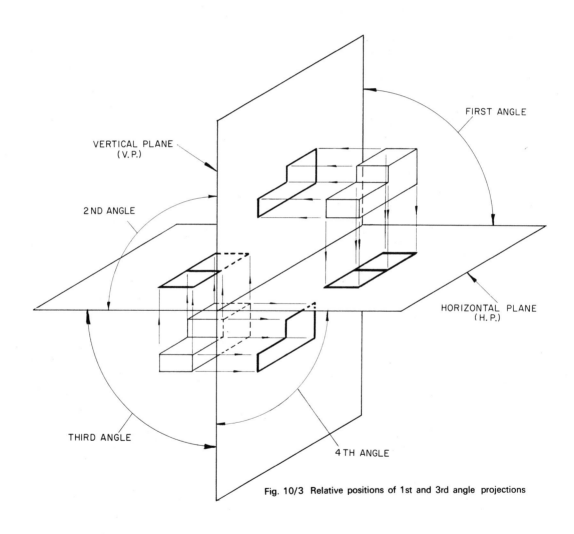

Fig. 10/3 Relative positions of 1st and 3rd angle projections

So far we have obtained only two views of the block, one on the V.P. and one on the H.P. With a complicated block this may not be enough. This problem is easily solved by introducing another plane. In this case it is a vertical plane and it will show a view of the end of the block and so, to distinguish it from the other vertical plane, it is called the end vertical plane (E.V.P.), and the original vertical plane is called the front vertical plane (F.V.P.).

The E.V.P. is hinged to the F.V.P. and when the views have been projected onto their planes, the three planes are unfolded and three views of the block are shown, Fig. 10/4.

The drawing on the F.V.P. is called the front elevation (F.E.), the drawing on the E.V.P. is called the end elevation (E.E.) and the drawing on the H.P. is called the plan. All three views are linked together: the plan is directly above the F.E.; the E.E. is horizontally in line with the F.E.; and the plan and the E.E. can be linked by drawing 45° projection lines. This is why orthographic projection is so important; it isn't just because several views of the same object can be drawn, *it is because the views are linked together*.

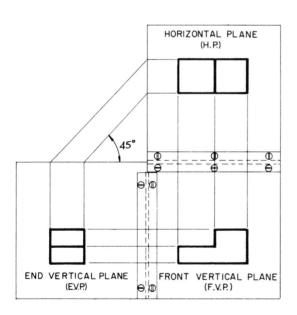

Fig. 10/4 3rd angle orthographic projection

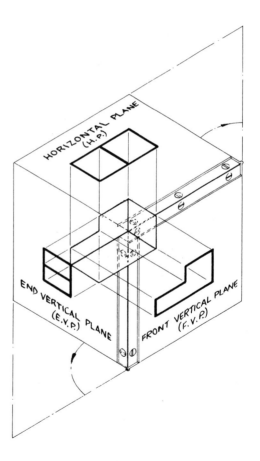

Fig. 10/4 showed three views of the block drawn in 3rd angle; Fig. 10/5 shows three views of the same block drawn in 1st angle.

In this case the F.E. is above the plan and to the left of the E.E. (compare this with 3rd angle). Once again, the E.E. and the plan can be linked by projection lines drawn at 45°.

The system of suspending the block between three planes and projecting views of the block onto these planes is the basic principle of orthographic projection and must be completely understood if one wishes to study this type of projection any further. This is done in Chapter 13.

The following system is somewhat easier to understand and will meet most of the readers needs.

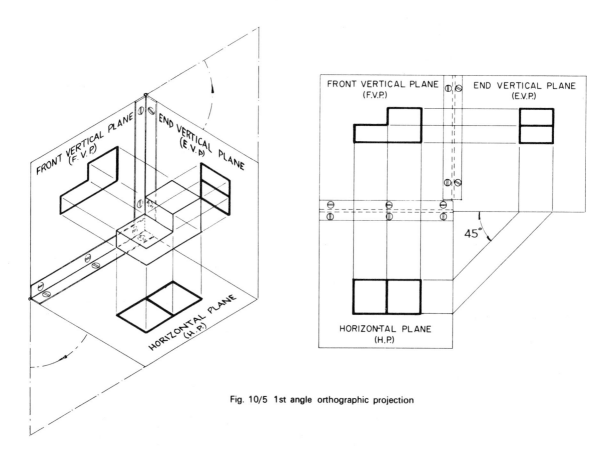

Fig. 10/5 1st angle orthographic projection

Fig. 10/6 shows the same shaped block drawn in 3rd angle projection. First, draw the view obtained by looking along the arrow marked F.E. This gives you the front elevation. Now look along the arrow marked E.E.₁ (which points from the left) and draw what you see to the left of the front elevation. This gives you an end elevation. Now look along the arrow marked E.E.₂ (which points from the right) and draw what you see to the right of the front elevation. This gives you another end elevation. Now look down onto the block, along the arrow marked 'plan' and draw what you see above the front elevation. This gives the plan and its exact position is determined by drawing lines from one of the end elevations at 45°.

Note that with 3rd angle projection, *what you see from the left you draw on the left, what you see from the right, you draw on the right, and what you see from above you draw above.*

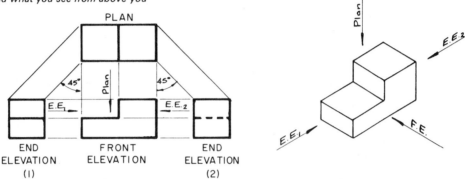

PLAN

45°   Plan   45°

E.E.₁   E.E.₂

END ELEVATION (I)   FRONT ELEVATION   END ELEVATION (2)

Plan   E.E.₂   E.E.₁   F.E.

Fig. 10/6 3rd angle orthographic projection

Fig. 10/7 shows the same block drawn in 1st angle projection. Again, first draw the view obtained by looking along the arrow marked F.E. This gives the front elevation. Now look along the arrow marked E.E.₁ (which points from the left) and draw what you see to the right of the front elevation. This gives you an end elevation. Now look along the arrow marked E.E.₂ (which points from the right) and draw what you see to the left of the front elevation. This gives you another end elevation. Now look down on the block, along the arrow marked 'plan' and draw what you see below the front elevation. This gives the plan and its exact position is determined by drawing lines from one of the end elevations at 45°.

Note that with 1st angle projection, *what you see from the left you draw on the right, what you see from the right you draw on the left, and what you see from above you draw below.*

Plan   E.E.₂   E.E.₁   F.E.

Plan

FRONT ELEVATION

E.E.₁   E.E.₂

END ELEVATION (2)   45°   45°   END ELEVATION (I)

PLAN

Fig. 10/7 1st angle orthographic projection

83

Auxiliary elevations and auxiliary plans

So far we have been able to draw four different views of the same block. In most engineering drawings these are sufficient but there are occasions when other views are necessary, perhaps to clarify a particular point. Fig. 10/8 shows two examples where a view other than a F.E. or an E.E. is needed to show very important features of a flanged pipe and a bracket.

AUXILIARY PLAN
SHOWING FACE
OF FLANGE

3RD ANGLE PROJECTION

AUXILIARY ELEVATION IN
DIRECTION OF ARROW

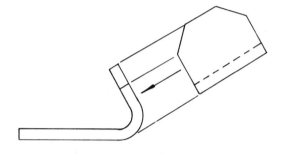

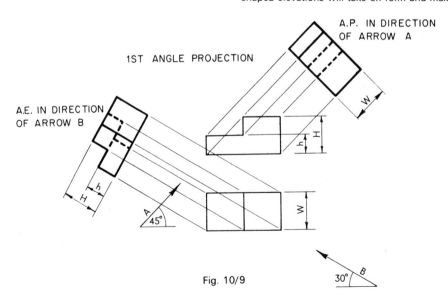

Fig. 10/8   These extra elevations are called auxiliary elevations or auxiliary plans.

Fig. 10/9 shows an auxiliary elevation (A.E.) and an auxiliary plan (A.P.) of the shaped block. One is projected from the plan at 30° and the other from the F.E. at 45°. Projection lines are drawn at those angles and the heights, H and h, are marked off on one A.E. and the width W on the other. Remember that we are dealing with a solid block, not flat shapes on flat paper. Try to imagine the block as a solid object and these rather odd-shaped elevations will take on form and make sense.

1ST ANGLE PROJECTION

A.P. IN DIRECTION
OF ARROW A

A.E. IN DIRECTION
OF ARROW B

Fig. 10/9

Fig. 10/10 shows two auxiliary plans of a more complicated block. In this case the base is tilted and therefore cannot be used to measure the heights as before. This is overcome by drawing a datum line. The heights of all the corners are measured from this datum. Note that on the auxiliary plans the datum is drawn at 90° to the projection lines.

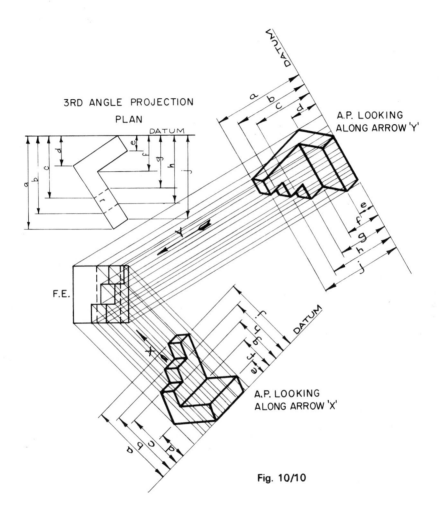

Fig. 10/10

If the outline contains circles or curves, the treatment is similar. Select some points on the curve and mark off their distances from some convenient datum. In Fig. 10/11 this gives dimensions *a*, *b*, *c*, *d*, *e* and *f*. The positions of these points are marked on the plan and they are projected onto the A.E. The dimensions *a* to *f* are then marked off on the A.E. and the points joined together with a neat freehand curve.

It is worth stating again the difference between 1st and 3rd angle projection, particularly if checked against the above examples. With 1st angle, if you look from one side of a view you draw what you see on the *other* side of that view. With 3rd angle, if you look from one side of a view you draw what you see on the *same* side of that view.

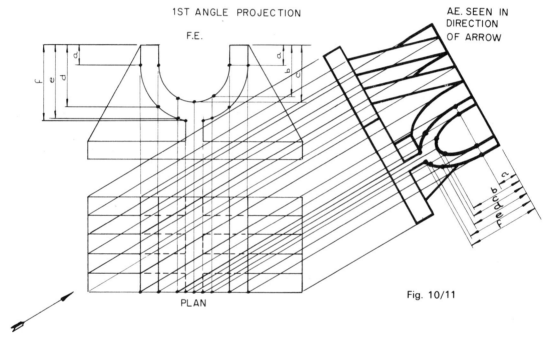

Fig. 10/11

Shown below are some of the more common solid geometric solids drawn in orthographic projection.

Prisms and Pyramids

Fig. 10/12 shows the following views of a rectangular prism, drawn in 1st angle projection with the prism tilted at 45° in the F.E.

A F.E. looking along Arrow A.
An E.E. looking along Arrow B.
A plan.
An A.P. showing the cross-sectional shape of the prism.

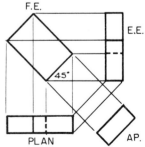

Fig. 10/12

Fig. 10/13 shows the following views of a square prism drawn in 3rd angle projection. The top of the prism has been cut obliquely at 30°.

A F.E. looking along Arrow A.
An E.E. looking along Arrow B.
A plan.
An A.P. projected from the F.E. at 30°.

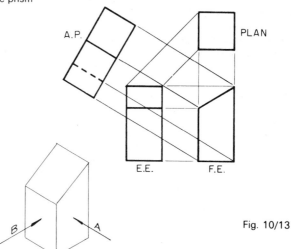

3RD ANGLE PROJECTION

A.P.

PLAN

E.E.   F.E.

Fig. 10/13

Fig. 10/14 shows the following views of a regular hexagonal prism, drawn in 3rd angle projection with the prism tilted at 30° in the F.E. The top of the prism has been cut obliquely at 45°.

A F.E. looking along Arrow A.
An E.E. looking along Arrow B.
A plan.

The first view that is drawn is the A.P. This is not in the instructions but without it the F.E. is very difficult to draw. Arrow A indicates that three sides of the hexagon are seen in the F.E. and the A.P. is constructed so that three sides are seen (rotate the hexagon through 30° in the A.E. and only two sides are seen). The A.P. is also used to find the width of the prism in the E.E.

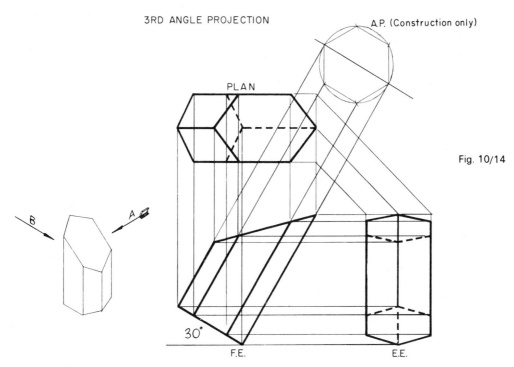

3RD ANGLE PROJECTION

A.P. (Construction only)

PLAN

Fig. 10/14

30°

F.E.   E.E.

87

Fig. 10/15 shows the following views of the frustum of a square pyramid drawn in 1st angle projection. The corners of the pyramid are numbered 1 to 4 for easy identification on each elevation.

A F.E. looking along the arrow.
An E.E. seen from the left of the F.E.
A plan.

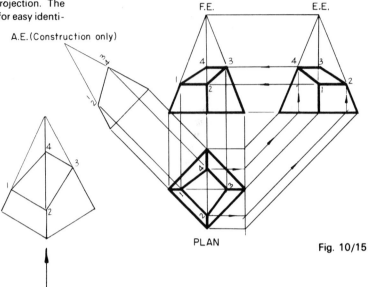

A.E. (Construction only)

F.E.   E.E.

PLAN

Fig. 10/15

With this type of problem it is wise initially to draw the required views as if the pyramid were complete. Once again it is necessary to draw an A.E. so that the oblique face can be drawn on the A.E. and then points 1, 2, 3 and 4 can be projected back onto the plan. Points 1 and 3 are then projected onto the F.E. and points 2 and 4 onto the E.E. Points 2 and 4 can be projected from the E.E. to the F.E. and points 1 and 3 from the F.E. to the E.E. Note that once the A.E. has been drawn it is possible to draw the oblique face on all three views without any further measuring.

Fig. 10/16 shows the following views of an octagonal pyramid drawn in 3rd angle projection. The pyramid is lying on its side.

A F.E. looking along the arrow.
An E.E. seen from the right of the F.E.
A plan.

To draw the pyramid lying on its side, first draw it standing upright and then tip it over. This is done with compasses as shown. If a plan of the pyramid standing upright is constructed, it makes it easier to find the positions of the corners of the pyramid in the plan when it has been tipped over.

3RD ANGLE PROJECTION

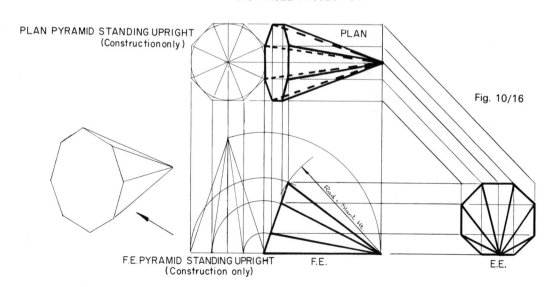

PLAN PYRAMID STANDING UPRIGHT
(Construction only)

PLAN

Fig. 10/16

Rad = Slant Ht.

F.E. PYRAMID STANDING UPRIGHT
(Construction only)

F.E.   E.E.

Fig. 10/17 shows the following views of a hexagonal pyramid drawn in 3rd angle projection. The top of the pyramid is cut at 45° and the bottom at 30°.

A F.E. seen in the direction of the arrow.

An E.E. seen from the right of the F.E.

A plan.

An A.E. projected from the plan at 30°.

As for Fig. 10/15, the pyramid is first drawn as if it were complete, on all four views. The lower cutting plane is then drawn on the F.E. The points where it crosses the corners are then projected across to the E.E. and up to the plan. The point where it crosses the centre corner on the F.E. cannot be projected straight to the plan and has to be projected via the E.E. (follow the arrows).

The upper cutting plane is then drawn on the E.E. and the points where it crosses the corners are projected across to the F.E. and up to the plan.

Most of these corners can be projected straight from the plan onto the A.E. The exceptions are the points on the centre corner and these (dimensions *a*, *b*, *c* and *d*) can be transferred from any convenient source, in this case the F.E.

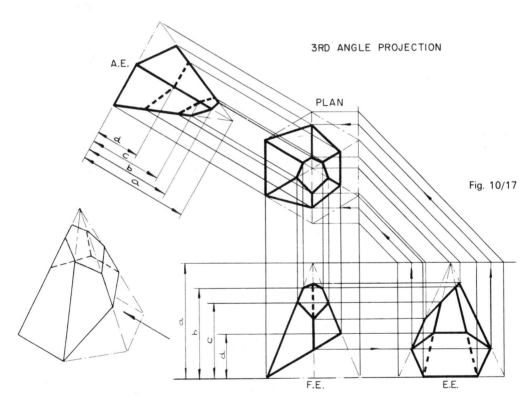

3RD ANGLE PROJECTION

Fig. 10/17

89

Cylinders and Cones

Fig. 10/18 shows the following views of a cylinder drawn in 1st angle projection.

A F.E. seen in the direction of the arrow.

A plan.

An A.P. projected from the F.E. at 45°.

If the plan is divided into a number of strips the width of the cylinder at any one of these strips can be measured. The exact positions of each of the strips can be projected onto the F.E. and then across to the A.P. The widths of the cylinder at each of the strips is transferred from the plan onto the A.P. with dividers, measured each side of the centre line (only one side is shown). The points are then joined together with a neat freehand curve.

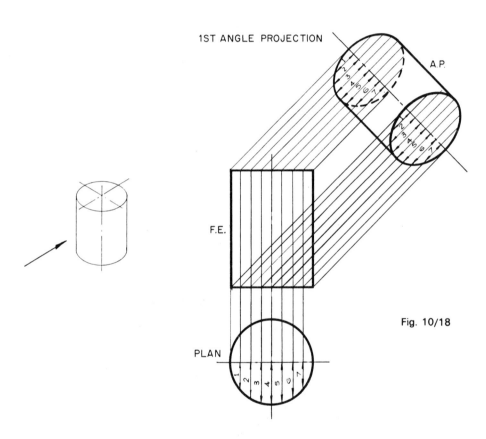

1ST ANGLE PROJECTION

A.P.

F.E.

Fig. 10/18

PLAN

Fig. 10/19 shows the following views of a cylinder drawn in 3rd angle projection. The cylinder is lying on its side and one end is cut off at 30° and the other end at 60°.

A F.E. seen in the direction of the arrow.

An E.E. seen from the left of the F.E.

A plan.

An A.P. projected from the plan at 60°.

The E.E. is divided into a number of strips. The strips are projected from the E.E. to the F.E. and up to the plan. They are also projected from the E.E. to the plan at 45°. The points where the projectors from the F.E. and the E.E. meet on the plan (at *a*, *b*, *c* and *d*, etc) give the outline of the two ellipses on the plan.

The outline of the ellipses on the A.P. are found by projecting the strips onto the A.P. and then transferring measurements 1, 2, 3, etc. from the E.E. to the A.P. with dividers.

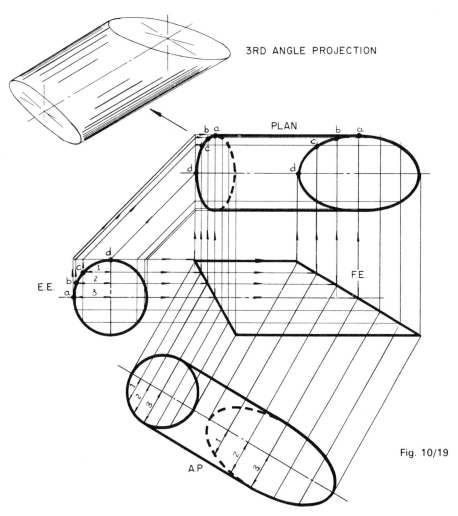

3RD ANGLE PROJECTION

PLAN

E.E.

F.E.

A.P

Fig. 10/19

Fig. 10/20 shows the following views of a curved cylinder drawn in 1st angle projection.

A F.E. seen in the direction of the arrow.

An E.E. seen from the left of the F.E.

A plan.

An A.P. projected from the F.E. at 45°.

This drawing uses a different method of plotting an A.P. from the previous two examples. Instead of being divided into strips, the cylinder is divided into 12 equal segments. These are marked on the walls of the cylinder as numbers, from 1 to 12. The ellipses formed on the A.P. are found by plotting the intersections of the projectors of numbers 1 to 12 from the F.E. and from a construction drawn in line with the A.P. The projectors intersect in 1', 2', 3' etc. Note that on the E.E. number 1 is at the top of the circle whilst on the construction (and hence on the A.P.) number 1 is on the right. This, of course, is what you should expect.

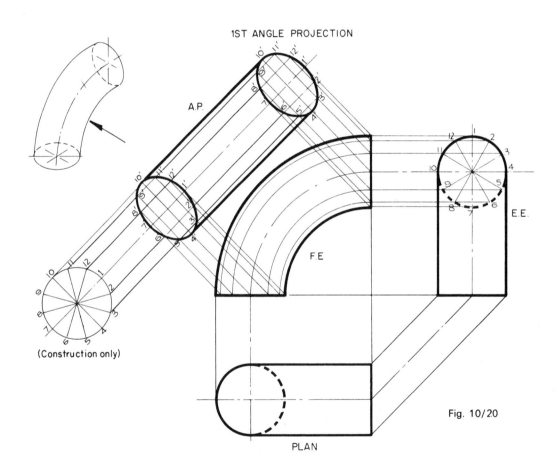

Fig. 10/20

92

Fig. 10/21 shows the following views of a cylinder drawn in 3rd angle projection. The base of the cylinder is cut obliquely at 30° and the cylinder is tilted at 60° in the F.E.

A F.E. seen in the direction of the arrow.
An E.E. seen from the left of the F.E.
A plan.

The cylinder is divided into twelve equal segments. This is done on a separate auxiliary elevation and plan which are constructed just for that purpose. The ellipses are found by plotting the intersections of the projectors from points 1 and 1, 2 and 2, 3 and 3, 4 and 4, etc.

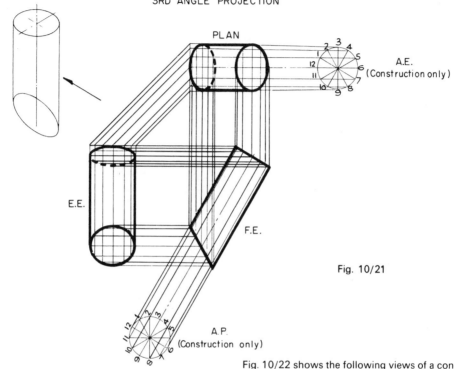

3RD ANGLE PROJECTION

PLAN

A.E.
(Construction only)

E.E.

F.E.

Fig. 10/21

A.P.
(Construction only)

1ST ANGLE PROJECTION

A.P.   E.E.   F.E.

PLAN

Fig. 10/22 shows the following views of a cone drawn in 1st angle projection.

A F.E. seen in the direction of the arrow.
An E.E. seen from the right of the F.E.
A plan.
An A.P. projected from the E.E. at 30°.

The plan is divided into strips. These strips are projected across to the E.E. and hence to the A.P. The width of the base of the cone on each of these strips is measured on the plan with dividers and transferred onto the A.P. The points are then joined with a neat freehand curve.

Fig. 10/22

93

Fig. 10/23 shows the following views of a cone drawn in 3rd angle projection. The cone is lying on its side.

A F.E. seen in the direction of the arrow.

An E.E. seen from the right of the F.E.

A plan.

The cone is first drawn standing upright and it is then tipped over to lie on its side. Instead of being divided into strips as before, it is divided into 12 equal segments. These are numbered from 1 to 12 on two constructions drawn in line with the F.E. and the E.E. The ellipses formed on the E.E. and the plan are found by plotting the intersections of the projectors from these constructions. They intersect in 1′, 2′, 3′, etc. on the E.E. and on the plan.

The geometry of cones is explored much more fully in the next chapter.

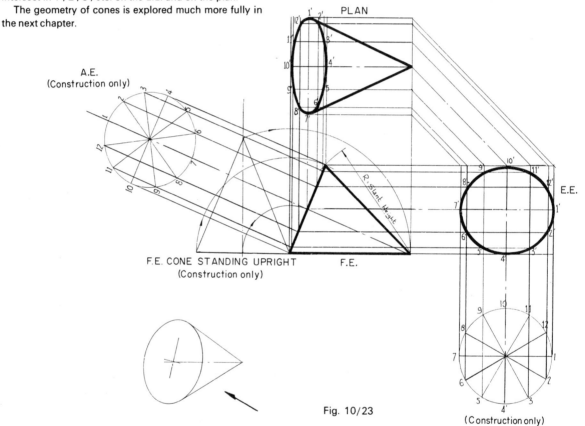

3RD ANGLE PROJECTION

PLAN

A.E.
(Construction only)

E.E.

F.E. CONE STANDING UPRIGHT
(Construction only)

F.E.

Fig. 10/23

(Construction only)

Sections

Suppose that you make a drawing of a box. You draw the box in orthographic projection and are pleased with the result. But someone comes along and says, quite reasonably, 'It's a good drawing but, after all, a box is only a container and you haven't shown what is inside the box; surely that is what is important'. And of course, he is right.

It is often vital to show what is inside an object as well as to show the outside. In orthographic projection, this is catered for with a section.

Fig. 10/24 shows two drawings of the same block, one drawn without a section and one drawn with a section. The upper drawing does not show clearly on any one of the orthographic views that the block is hollow. On the lower left isometric view, the block has been cut in half and it is immediately obvious that the block is hollow. The lower right view shows the cut block drawn in orthographic projection. Again, it is much easier to see that the block is hollow.

Note carefully the following rules:

1. The sectioned E.E. is drawn with half of the block missing *but none of the other views are affected*. They keep their normal full outline.
2. The point where the section is made is denoted by a cutting plane. This is drawn with a thin chain dot line which is thickened where it changes direction and for a short distance at the end. The arrows point in the direction that the section is projected.
3. Where the cutting plane cuts through solid material, the material is hatched at 45°.
4. When a section is projected, the remaining visible features which can be seen on the other side of the cutting plane are also drawn on the section.
5. It is not usual to draw hidden detail on a section.

There are many rules about sectioning but most of them apply to engineering drawing, rather than geometric drawing. For this reason they are found in the second part of this book.

3RD ANGLE PROJECTION

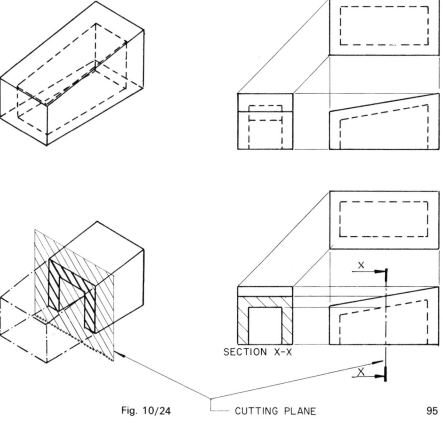

SECTION X-X

Fig. 10/24   CUTTING PLANE

95

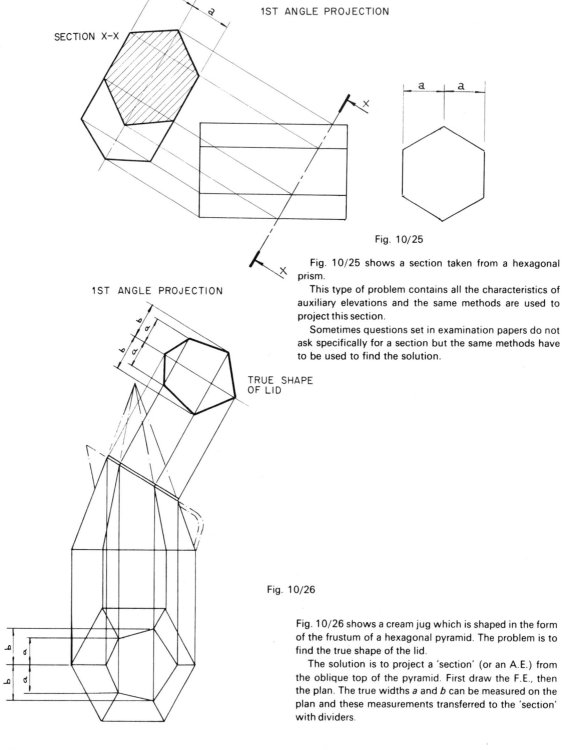

1ST ANGLE PROJECTION

SECTION X-X

Fig. 10/25

Fig. 10/25 shows a section taken from a hexagonal prism.

This type of problem contains all the characteristics of auxiliary elevations and the same methods are used to project this section.

Sometimes questions set in examination papers do not ask specifically for a section but the same methods have to be used to find the solution.

1ST ANGLE PROJECTION

TRUE SHAPE
OF LID

Fig. 10/26

Fig. 10/26 shows a cream jug which is shaped in the form of the frustum of a hexagonal pyramid. The problem is to find the true shape of the lid.

The solution is to project a 'section' (or an A.E.) from the oblique top of the pyramid. First draw the F.E., then the plan. The true widths a and b can be measured on the plan and these measurements transferred to the 'section' with dividers.

Fig. 10/27 shows a section projected from a piece of thick wall tubing that is attached to a rectangular base. Apart from the F.E. a plan must be drawn. Points around the circumference of the circles on the plan are selected and these points are projected down to the section plane and then across onto the section itself. The centre line is a convenient datum and so the distance from each point to the centre line is measured with dividers and transferred to its corresponding projection line on the section. For clarity, only seven points are shown but, of course, all the points round the circumference would have to be measured and transferred.

There are two points to note from this drawing. Firstly, the section hatching is not drawn at 45°. This is because the hatching would then be parallel to a large part of the outline. In this type of case, an angle other than 45° can be adopted. Secondly, note that hatching is done only where the section (or cutting) plane actually cuts through solid material.

3RD ANGLE PROJECTION

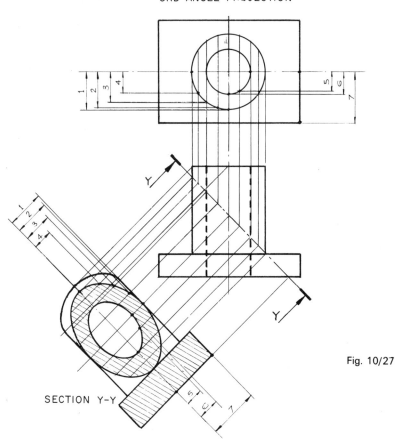

SECTION Y-Y

Fig. 10/27

Finally, Fig. 10/28 shows a section through a socket spanner. One end has a hexagonal recess to fit a nut or bolt head and the other has a square recess to accommodate a tommy bar. The circular profile is again split up so that points along its circumference can be measured and the measurements transferred onto the section via the section plane. The points where the section plane crosses the hexagonal and square sockets are projected onto the section and any measurements that have to be made are found by projecting back to one of the end elevations. These are marked *a*, *b*, *c* and *d*. The measurements for the outside profile are marked from 1 to 8.

For clarity only half of the measurements are shown. The other half is dealt with in the same way.

Conic sections are dealt with in the next chapter.

Sections applied to engineering drawings are dealt with in Part 2 of this book.

3RD ANGLE PROJECTION

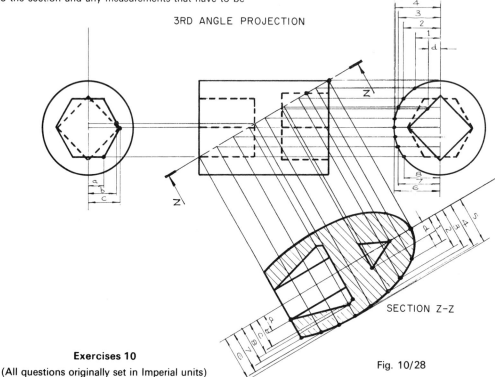

SECTION Z–Z

Fig. 10/28

Exercises 10

(All questions originally set in Imperial units)

1. Fig. 1. shows the elevation of a 20 mm square prism 50 mm long resting with one of its corners on the horizontal plane. Draw, full size, the following views and show all the hidden detail.
   (a) The given elevation.
   (b) An end elevation looking in the direction of arrow E.
   (c) A plan projected beneath view (a).
   North Western Secondary School Examinations Board

2. Fig. 2 shows the front elevation and an isometric sketch of a sheet metal footlight reflector for a puppet theatre.
   Draw, full size, the given front elevation and from it project the plan and end view of the reflector.
   Draw, also, the true shape of the surface marked ABCD.
   The thickness of the metal can be ignored.
   West Midlands Examinations Board

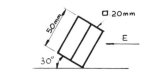

Fig. 1

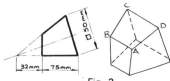

Fig. 2

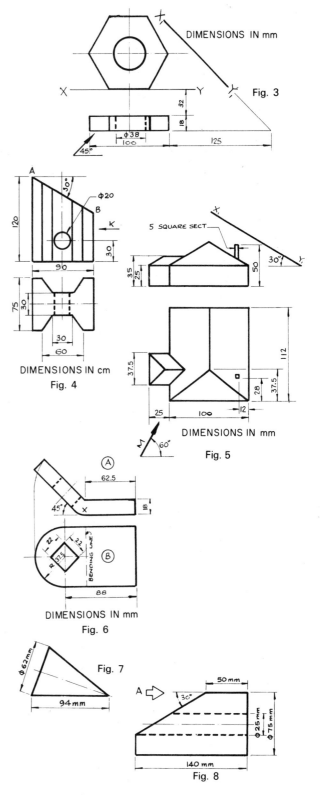

3. The Plan and Elevation of a hexagonal distance piece are shown in Fig. 3. Draw these views, full size, and project an auxiliary elevation on X_1Y_1. Hidden details are not to be shown.
Associated Lancashire Schools Examining Board

DIMENSIONS IN mm

Fig. 3

4. Fig. 4 shows details of a cast concrete block. To a scale of 10 mm = 100 mm draw the following:
(a) The two given views.
(b) An end elevation looking in the direction of the arrow K.
(c) The true shape of the sloping surface AB.
Metropolitan Regional Examinations Board

DIMENSIONS IN cm
Fig. 4

5. Two views are shown of a bungalow which has been made as a model (Fig. 5). To obtain a better impression of its design a view in the direction of arrow M is required. Draw full size the following:
(a) The two given views.
(b) An auxiliary elevation, projected from the plan, in the direction of arrow M.
Metropolitan Regional Examinations Board

DIMENSIONS IN mm
Fig. 5

6. The front view of Fig. 6 (drawing A) is of a piece of metal with the left-hand portion bent upwards at an angle of 45 degrees as shown. The bottom drawing is the plan view of the metal before it was bent.
Draw, *full size*,
(a) the given front view;
(b) the plan of the piece of metal *after it has been bent up*. The curve X can be ignored.
South-East Regional Examinations Board

DIMENSIONS IN mm
Fig. 6

7. An elevation of a cone lying on its side is given in Fig. 7. Copy the given elevation full size and from it project the plan and end view of the cone.
West Midlands Examinations Board

Fig. 7

8. An elevation of a machined part is given in Fig. 8. Draw the following views *full size*:
(a) the elevation as shown;
(b) an end view as seen in the direction of arrow 'A';
(c) the true shape of the sloping surface.
South-East Regional Examinations Board

Fig. 8

99

9. Fig. 9 shows a church steeple which is square in plan. to a scale of 1:100 draw
   (a) a Front Elevation;
   (b) a Sectional Plan on a cutting plane 2600 mm above the centre of the clock.
   Ignore the chain lines.
   South-East Regional Examinations Board

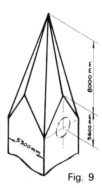

Fig. 9

10. Draw the two elevations of the machined section shown in Fig. 10 and add a plan in the direction of the arrow P showing hidden lines. Scale full size.
    Oxford and Cambridge Schools Examination Board

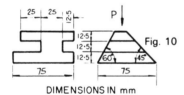

Fig. 10

DIMENSIONS IN mm

11. Fig. 11 shows the elevation of a right hexagonal pyramid of base edges 35 mm and vertical height 80 mm. Draw the given elevation and the true shape of the surface contained in the section plane X–X.
    University of London School Examinations

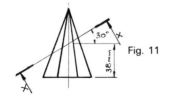

Fig. 11

12. Three views of a model of a steam turbine are shown in Fig. 12.
    Draw, to a scale of half-full size, showing all hidden detail:
    (a) the given plan and front elevation, and
    (b) an auxiliary plan in the direction of 'KK'.
    Associated Examining Board

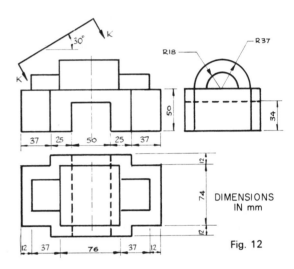

DIMENSIONS IN mm

Fig. 12

13. Details of an angle bracket are shown in Fig. 13. Draw the two given views, the view as seen from the direction of arrow A and an elevation as seen from the left of the front elevation. Hidden detail need not be shown. Scale: full size; use first angle orthographic projection.
    Oxford Local Examinations

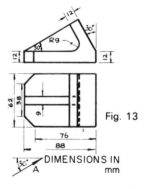

Fig. 13

DIMENSIONS IN mm

14. The plan and elevation of a special angle bracket are shown in Fig. 14.
    (a) Draw, full-size, the given views and project an auxiliary plan on the ground line X_1–Y_1.
    (b) Using the auxiliary plan in (a) above, project an auxiliary elevation on the ground line X_2–Y_2.
    All hidden detail to be shown.
    Associated Examining Board

16. Fig. 16 illustrates an elevation of an extension leg for a socket spanner.
    Draw, to a scale of 4 : 1 and in first angle orthographic projection:
    (a) the given elevation;
    (b) a plan;
    (c) the true shape of the section at the cutting plane XX.
    Oxford Local Examinations

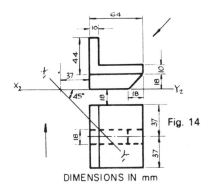

Fig. 14

DIMENSIONS IN mm

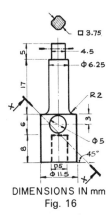

DIMENSIONS IN mm

Fig. 16

15. Two views of a pivot block are shown in Fig. 15. Draw the given views, and produce an elevation on XY as seen when looking in the direction of arrow A. Hidden edges are to be shown.
    Cambridge Local Examinations

17. Fig. 17 shows an elevation and a plan of a casting of a corner cramp in which the shape is symmetrical about AA.
    Draw a second elevation looking in the direction of arrow B.
    Draw a sectional elevation on the cutting plane AA.
    Hidden detail is required on all views.
    Southern Universities' Joint Board

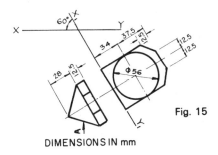

Fig. 15

DIMENSIONS IN mm

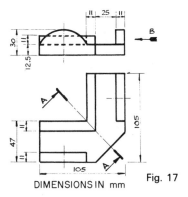

DIMENSIONS IN mm

Fig. 17

11

Conic sections—the ellipse, the parabola, the hyperbola

Fig. 11/1 shows the five sections that can be obtained from a cone. The triangle and the circle have been discussed in earlier chapters; this chapter looks at the remaining three sections, the ellipse, the parabola and the hyperbola. These are three very important curves. The ellipse can vary in shape from almost a circle to almost a straight line and is often used in designs because of its pleasing shape. The parabola can be seen in the shape of electric fire reflectors, radar dishes and the main cable of suspension bridges. Both the parabola and the hyperbola are much used in civil engineering. The immense strength of structures which are parabolic or hyperbolic in shape has led to their use in structures made of pre-cast concrete and where large unsupported ceilings are needed.

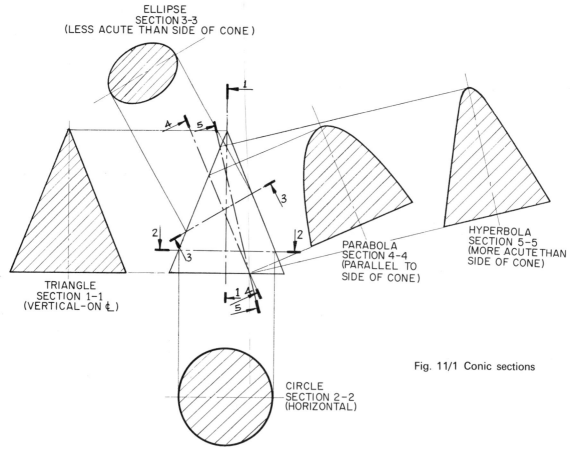

1ST ANGLE PROJECTION

ELLIPSE
SECTION 3-3
(LESS ACUTE THAN SIDE OF CONE)

HYPERBOLA
SECTION 5-5
(MORE ACUTE THAN
SIDE OF CONE)

PARABOLA
SECTION 4-4
(PARALLEL TO
SIDE OF CONE)

TRIANGLE
SECTION 1-1
(VERTICAL-ON ℄)

CIRCLE
SECTION 2-2
(HORIZONTAL)

Fig. 11/1 Conic sections

The height of the cone and the base diameter, together with the angle of the section relative to the side of the cone, are the factors which govern the relative shape of any ellipses, parabolas or hyperbolas. There are an infinite number of these curves and, given eternity and the inclination to do so, you could construct them all by taking sections from cones. There are other ways of constructing these curves and this chapter, as well as showing how to obtain them by plotting them from cones, shows some other equally important methods of construction.

THE ELLIPSE

Fig. 11/2 shows an ellipse with the important features labelled.

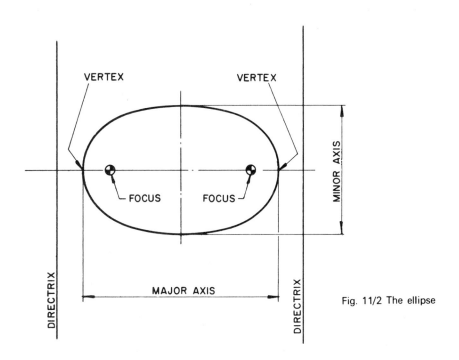

Fig. 11/2 The ellipse

The ellipse as a conic section

Fig. 11/3 shows in detail how to project an ellipse as a section of a cone. The shape across X–X is an ellipse.

First draw the F.E., the E.E. and the plan of the complete cone. Divide the plan into 12 equal sectors with a 60° set square. Project these sectors onto the F.E. and the E.E. where they appear as lines drawn on the surface of the cone from the base to the apex. The points where these lines cross X–X can be easily projected across to the E.E.

and down to the plan to give the shape of X–X on these elevations. The point on the centre line must be projected onto the plan via the E.E. (follow the arrows).

The shape of X–X on the plan is not the true shape since, in the plan, X–X is sloping down into the page. However, the widths of the points, measured from the centre line, are true lengths and can be transferred from the plan to the auxiliary view with dividers to give the true shape across X–X. This is the ellipse.

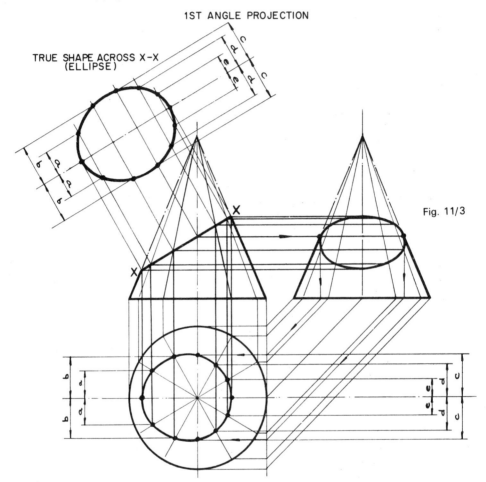

1ST ANGLE PROJECTION

TRUE SHAPE ACROSS X–X (ELLIPSE)

Fig. 11/3

ANGLE OF X–X TO BASE IS LESS THAN THE SIDE OF THE CONE

The ellipse as a locus
Definition

An ellipse is the locus of a point which moves so that its distance from a fixed point (called the focus) bears a constant ratio, always less than 1, to its perpendicular distance from a straight line (called the directrix). An ellipse has two foci and two directrices.

Fig. 11/4 shows how to draw an ellipse given the relative positions of the focus and the directrix, and the eccentricity. In this case the focus and the directrix are 20 mm apart and the eccentricity is $\frac{3}{4}$.

The first point to plot is the one that lies between the focus and the directrix. This is done by dividing DF in the

$\frac{4}{3} \times 50$

$\frac{4}{3} \times 40$

$\frac{4}{3} \times 30$

$\frac{4}{3} \times 20$

$\frac{4}{3} \times 10$

DIRECTRIX

R10 R20 R30 R40 R50

D

F (FOCUS)

P

Fig. 11/4

20

$FP = \frac{3}{4} DP, \therefore FP = \frac{3}{4}(FP+20)$, Hence $FP = 60$

ECCENTRICITY $\frac{3}{4}$

DIMENSIONS IN mm

same ratio as the eccentricity, 4 : 3. The other end of the ellipse, point P, is found by working out the simple algebraic sum shown on Fig. 11/4.

The condition for the locus is that it is always $\frac{3}{4}$ as far from the focus as it is from the directrix. It is therefore $\frac{4}{3}$ as far from the directrix as it is from the focus. Thus, if the point is 30 mm from F, it is $\frac{40}{3}$ mm from the directrix; if the point is 20 mm from F it is $\frac{4}{3} \times 20$ mm from the directrix; if the point is 30 mm from F, it is $\frac{3}{4} \times 30$ mm from the directrix. This is continued for as many points as may be necessary to draw an accurate curve. The intersections of radii drawn from F and lines drawn parallel to the directrix, their distance from the directrix being proportional to the radii, give the outline of the ellipse. These points are joined together with a neat freehand curve.

STAGE 4

STAGE 1

R = $\frac{1}{2}$ Maj. Axis

R = $\frac{1}{2}$ Min. Axis

STAGE 3

STAGE 2

Fig. 11/5

To construct an ellipse by concentric circles

We now come to the first of three simple methods of constructing an ellipse. All three methods need only two pieces of information for the construction—the lengths of the major and minor axes.

Stage 1. Draw two concentric circles, radii equal to $\frac{1}{2}$ major and $\frac{1}{2}$ minor axes.

Stage 2. Divide the circle into a number of sectors. If the ellipse is not too large, twelve will suffice as shown in Fig. 11/5.

Stage 3. Where the sector lines cross the smaller circle, draw horizontal lines towards the larger circle. Where the sector lines cross the larger circle, draw vertical lines to meet the horizontal lines.

Stage 4. Draw a neat curve through the intersections.

To construct an ellipse in a rectangle
Stage 1. Draw a rectangle, length and breadth equal to the major and minor axes.
Stage 2. Divide the two shorter sides of the rectangle into the same *even* number of equal parts. Divide the major axis into the same number of equal parts.
Stage 3. From the points where the minor axis crosses the edge of the rectangle, draw intersecting lines as shown in Fig. 11/6.
Stage 4. Draw a neat curve through the intersections.

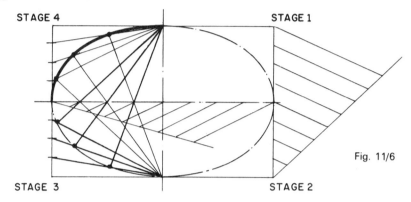

STAGE 4 STAGE 1

STAGE 3 STAGE 2

Fig. 11/6

To construct an ellipse with a trammel
A trammel is a piece of stiff paper or card with a straight edge. In this case, mark the trammel with a pencil so that half the major and minor axes are marked from the same point P. Keep B on the minor axis, A on the major axis and slide the trammel, marking at frequent intervals the position of P. Fig. 11/7 shows the trammel in position for plotting the top half of the ellipse; to plot the bottom half, A stays on the major axis and B goes above the major axis, still on the minor axis.

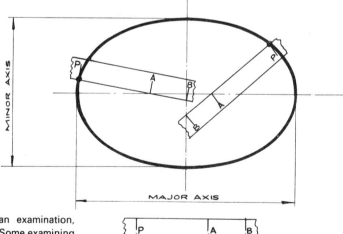

MINOR AXIS

MAJOR AXIS

Before using this construction in an examination, check the syllabus to see if it is allowed. Some examining boards do not allow an ellipse to be constructed with a trammel. If this is the case, use the concentric circles or rectangle method.

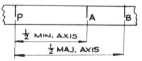

P A B

½ MIN. AXIS
½ MAJ. AXIS

Fig. 11/7

To find the foci, the normal and the tangent of an ellipse (Fig. 11/8)

The foci. With compasses set at a radius of ½ major axis, centre at the point where the minor axis crosses the top (or the bottom) of the ellipse, strike an arc to cut the major axis twice. These are the foci.

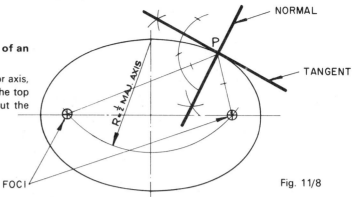

Fig. 11/8

The normal at any point P. Draw two lines from P, one to each focus and bisect the angle thus formed. This bisector is a normal to the ellipse.

The tangent at any point P. Since the tangent and normal are perpendicular to each other by definition, construct the normal and erect a perpendicular to it from P. This perpendicular is the tangent.

THE PARABOLA
The parabola as a conic section
The method used for finding the ellipse in Fig. 11/3 can be adapted for finding a parabolic section. However, the method shown below is much better because it allows for many more points to be plotted. In Fig. 11/9 the shape across Y–Y is a parabola.

First divide Y–Y into a number of equal parts, in this

case 6. The radius of the cone at each of the seven spaced points is projected on to the plan and circles are drawn. Each of the points must lie on its respective circle. The exact position of each point is found by projecting it onto the plan until it meets its circle. The points can then be joined together on the plan with a neat curve.

The E.E. is completed by plotting the intersection of the projectors of each point from the F.E. and the plan.

Neither the E.E. nor the plan show the true shape of Y–Y since, in both views, Y–Y is sloping into the paper. The only way to find the true shape of Y–Y is to project a view at right angles to it. The width of each point, measured from the centre line, can be transferred from the plan as shown.

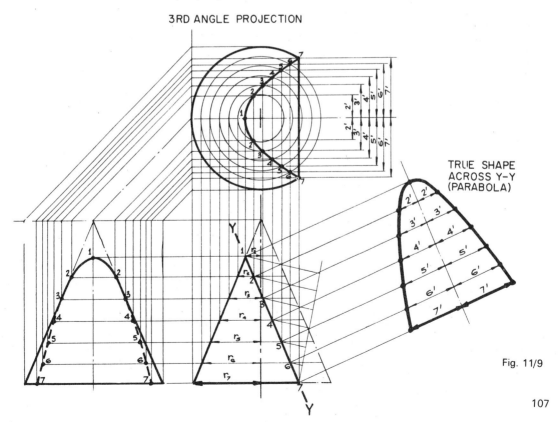

3RD ANGLE PROJECTION

TRUE SHAPE ACROSS Y–Y (PARABOLA)

Fig. 11/9

Y–Y IS PARALLEL TO THE SIDE OF THE CONE

107

The parabola as a locus

Definition

A parabola is the locus of a point which moves so that its distance from a fixed point (called the focus) bears a constant ratio of 1 to its perpendicular distance from a straight line (called the directrix).

Fig. 11/10 shows how to draw a parabola given the relative positions of the focus and the directrix. In this case the focus and directrix are 20 mm apart.

The first point to plot is the one that lies between the focus and the directrix. By definition it is the same distance, 10 mm, from both.

The condition for the locus is that it is always the same distance from the focus as it is from the directrix. The parabola is therefore found by plotting the intersections of radii 15 mm, 20 mm, 30 mm, etc., centre on the focus, with lines drawn parallel to the directrix at distances 15 mm, 20 mm, 30 mm, etc.

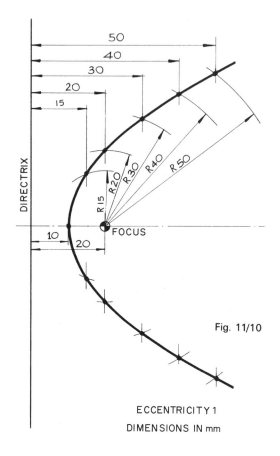

ECCENTRICITY 1

DIMENSIONS IN mm

Fig. 11/10

The parabola within a rectangle

This is a very simple construction and is shown in Fig. 11/11. The method should be obvious.

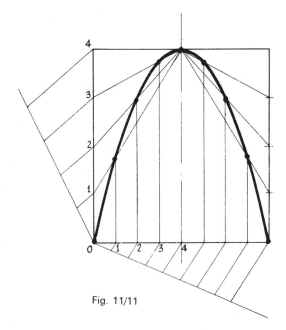

Fig. 11/11

108

To find the focus of a parabola and the tangent at a point P

Select a point R on the axis which is obviously further from V than the focus will be. From R erect a perpendicular and mark off RS = 2VR.

Join SV; this cuts the parabola in T.

From T drop a perpendicular to meet the axis in F.

F is the focus.

To draw the tangent at P, join FP and draw PQ parallel to the axis.

The bisector of FP̂Q is the tangent.

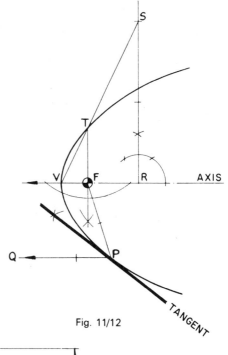

Fig. 11/12

THE HYPERBOLA
The hyperbola as a conic section

The method is identical to that used for finding the parabolic section in Fig. 11/9. The construction, Fig. 11/13, can be followed from the instructions for that figure.

3RD ANGLE PROJECTION

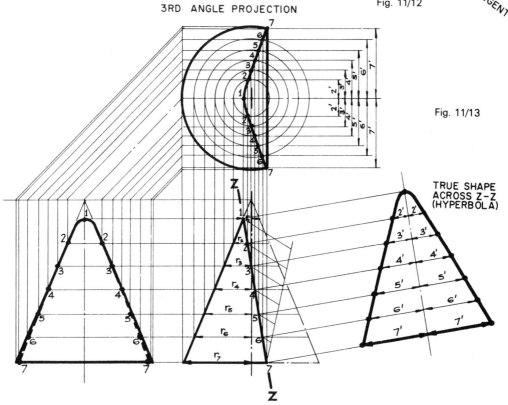

Fig. 11/13

TRUE SHAPE ACROSS Z–Z (HYPERBOLA)

ANGLE OF Z–Z TO THE BASE IS GREATER THAN THE SIDE OF THE CONE

109

The hyperbola as a locus

Definition

A hyperbola is the locus of a point which moves so that its distance from a fixed point (called the focus) bears a constant ratio, always greater than 1, to its perpendicular distance from a straight line (called the directrix).

Fig. 11/14 shows how to draw a hyperbola given the relative positions of the focus and the directrix (in this case 20 mm) and the eccentricity (3/2).

The first point to plot is the one that lies between the focus and the directrix. This is done by dividing the distance between them in the same ratio as the eccentricity, 3 : 2.

The condition for the locus is that it is always $\frac{2}{3}$ as far from the directrix as it is from the focus. Thus, if the point is 15 mm from the focus, it is $\frac{2}{3} \times 15$ mm from the directrix; if it is 20 mm from the focus, it is $\frac{2}{3} \times 20$ mm from the directrix. This is continued for as many points as may be required.

There are constructions for the normal and tangent to a hyperbola but they introduce additional features which are beyond the scope of this book.

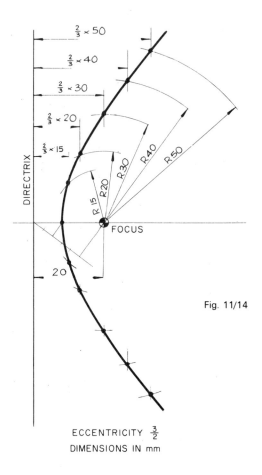

Fig. 11/14

ECCENTRICITY $\frac{3}{2}$

DIMENSIONS IN mm

Exercises 11

(All questions originally set in Imperial units)

1. Fig. 1 is the frustum of a right cone. Draw this elevation and a plan. Draw the true shape of the face AB.

Southern Regional Examinations Board

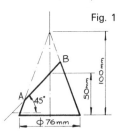

Fig. 1

2. Fig. 2 shows a point P which moves so that the sum of the distance from P to two fixed points, 100 mm apart, is constant and equal to 125 mm. Plot the path of the point P. Name the curve and the given fixed points.

Associated Lancashire Schools Examining Board

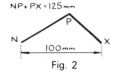

NP+ PX = 125 mm

Fig. 2

3. Fig. 3 shows the loud-speaker grill of a car radio. The grill is rectangular with an elliptical hole. Draw the grill, full size, showing the construction of the ellipse clearly.

West Midlands Examinations Board

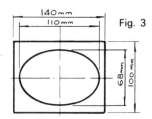

Fig. 3

4. Fig. 4 shows an elliptical fish-pond for a small garden. The ellipse is 1440 mm long and 720 mm wide. Using a scale of $\frac{1}{12}$, draw a true elliptical shape of the pond. (Do not draw the surrounding stones.)
All construction must be shown.
If a paper trammel is used, an accurate drawing of it must be made.
East Anglian Examinations Board

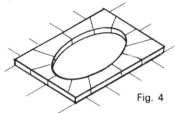

Fig. 4

5. Fig. 5 shows a section, based on an ellipse, for a handrail which requires cutting to form a bend so that the horizontal overall distance is increased from 112 mm to 125 mm.
Construct the given figures and show the tangent construction at P and P$_1$.
Show the true shape of the cut when the horizontal distance is increased from 112 mm to 125 mm.
Southern Universities' Joint Board (See Ch. 7 for information not in Ch. 11)

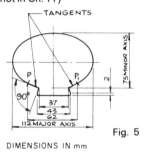

Fig. 5

DIMENSIONS IN mm

6. Fig. 6 shows the *upper half* of the section of a small headlamp. The casing is in the form of a semi-ellipse. F is the focal point. The reflector section is in the form of a parabola.
Part 1. Draw, *full size,* the complete semi-ellipse.
Part 2. Draw, *full size,* the complete parabola inside the semi-ellipse.
Southern Regional Examinations Board

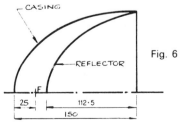

Fig. 6

DIMENSIONS IN mm

7. A point moves in a plane in such a way that its distance from a fixed point is equal to its shortest distance from a fixed straight line.
Plot the locus of the moving point when the fixed point is 44 mm from the fixed line. The maximum distance of the moving point is 125 mm from the fixed point.
State the name of the locus, the fixed point, and the fixed line.
Associated Examining Board

8. A piece of wire is bent into the form of a parabola. It fits into a rectangle which has a base length of 125 mm and a height of 100 mm. The open ends of the wire are 125 mm apart. By means of a single line, show the true shape of the wire.
Cambridge Local Examinations

9. An arch has a span of 40 m and a central rise of 13 m and the centre line is an arc of a parabola. Draw the centre line of the arch to a scale of 10 mm = 20 m.
Oxford and Cambridge Schools Examination Board

10. A cone, vertical height 100 mm, base 75 mm diameter, is cut by a plane parallel to its axis and 12 mm from it. Draw the necessary views to show the true shape of the section and state the name of it.
Oxford and Cambridge Schools Examination Board

11. Fig. 7 shows a right cone cut by a plane X–X. Draw the given view and project an elevation seen from the left of the given view.

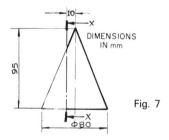

Fig. 7

12. Draw the conic having an eccentricity of $\frac{3}{4}$ and a focus which is 38 mm from the directrix. State the name of this curve.
Associated Examining Board

12

Intersection of regular solids

When two solids interpenetrate, a line of intersection is formed. It is sometimes necessary to know the exact shape of this line, usually so that an accurate development of either or both of the solids can be drawn. This chapter shows the lines of intersection formed when some of the simpler geometric solids interpenetrate.

Two dissimilar square prisms meeting at right angles (Fig. 12/1)
The E.E. shows where corners 1 and 3 meet the larger prism and these are projected across to the F.E. The plan shows where corners 2 and 4 meet the larger prism and this is projected up to the F.E.

Two dissimilar square prisms meeting at an angle (Fig. 12/2)
The F.E. shows where corners 1 and 3 meet the larger prism. The plan shows where corners 2 and 4 meet the larger prism and this is projected down to the F.E.

1ST ANGLE PROJECTION

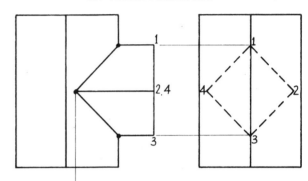

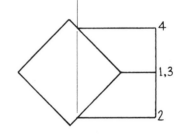

Fig. 12/1

3RD ANGLE PROJECTION

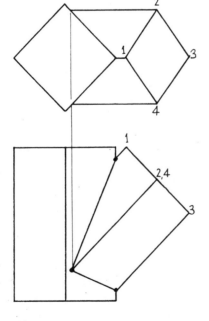

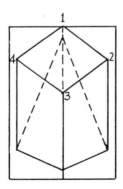

Fig. 12/2

A hexagonal prism meeting a square prism at right angles (Fig. 12/3)

The plan shows where all the corners of the hexagonal prism meet the square prism. These are projected down to the F.E. to meet the projectors from the same corners on the E.E.

Fig. 12/3

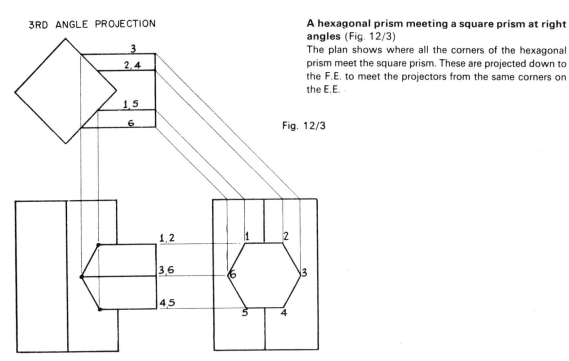

1ST ANGLE PROJECTION

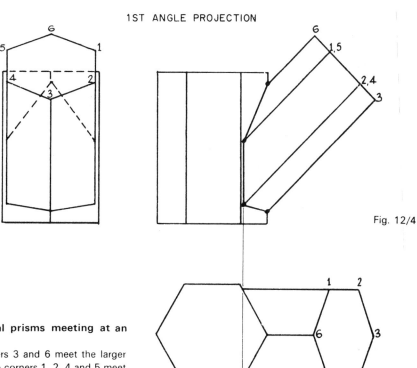

Fig. 12/4

Two dissimilar hexagonal prisms meeting at an angle (Fig. 12/4)

The F.E. shows where corners 3 and 6 meet the larger prism. The plan shows where corners 1, 2, 4 and 5 meet the larger prism and these are projected up to the F.E.

A hexagonal prism meeting an octagonal prism at an angle, their centres not being in the same vertical plane (Fig. 12/5)

The F.E. shows where corners 3 and 6 meet the octagonal prism. The plan shows where corners 1, 2, 4 and 5 meet the octagonal prism and these are projected down to the F.E.

The sides of the hexagonal prism between corners 3–4 and 5–6 meet two sides of the octagonal prism. The change of shape occurs at points *a* and *b*. The position of *a* and *b* on the F.E. (and then across to the E.E.) is found by projecting down to the F.E. via the end of the hexagonal prism (follow the arrows). The intersection on the F.E. can then be completed.

3RD ANGLE PROJECTION

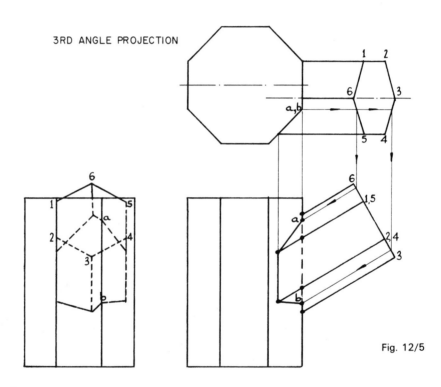

Fig. 12/5

114

A square prism meeting a square pyramid at right angles (Fig. 12/6)

The E.E. shows where corners 1 and 3 meet the pyramid. These are projected across to the F.E.

Corners 2 and 4 are not quite so obvious. The pictorial view shows how these corners meet the pyramid. If the pyramid was cut across X–X, the section of the pyramid resulting would be square, and points 2 and 4 would lie on this square. It isn't necessary to make a complete, shaded section on your drawing but it is necessary to draw the square on the plan. Since points 2 and 4 lie on this square it is simple to find their exact position. Project corners 2 and 4 from the E.E. onto the plan. The points where these projectors meet the square are the exact positions of the intersections of corners 2 and 4 with the pyramid.

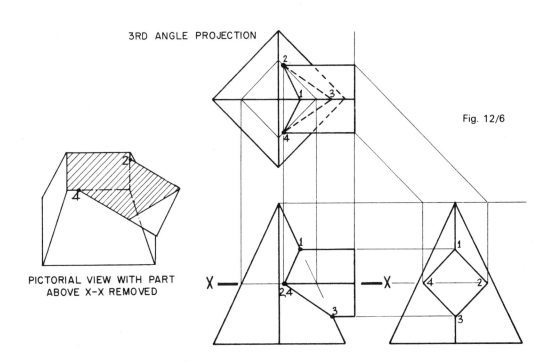

3RD ANGLE PROJECTION

Fig. 12/6

PICTORIAL VIEW WITH PART
ABOVE X-X REMOVED

115

A square pyramid and a hexagonal prism meeting at an angle (Fig. 12/7)

The F.E. shows where corners 1 and 4 meet the pyramid.

Corners 2 and 6 lie on the same plane X–X. If this plane is marked on the plan view of the pyramid (follow the arrows) it results in the line X–X–X. Corners 2 and 6 lie on this plane; their exact positions are as shown.

Corners 3 and 5 lie on the same plane Y–Y. On the plan view this plane is seen as the line Y–Y–Y (follow the arrows). Corners 3 and 5 lie on this plane; their exact positions are as shown.

1ST ANGLE PROJECTION

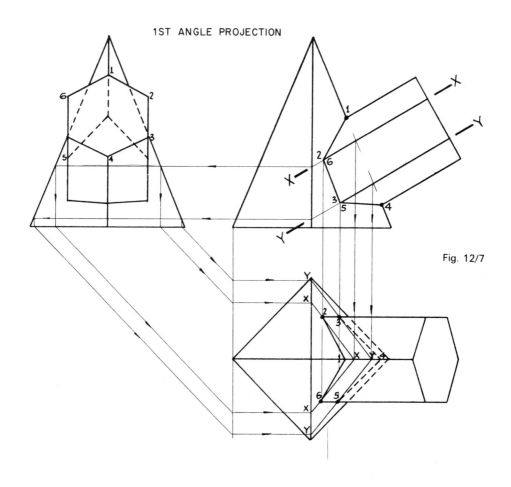

Fig. 12/7

116

Two dissimilar cylinders meeting at right angles
(Fig. 12/8)

The smaller cylinder is divided into 12 equal sectors on the F.E. and on the plan (the E.E. shows how these are arranged round the cylinder).

The plan shows where these sectors meet the larger cylinder and these intersections are projected down to the F.E. to meet their corresponding sector at 1', 2', 3', etc.

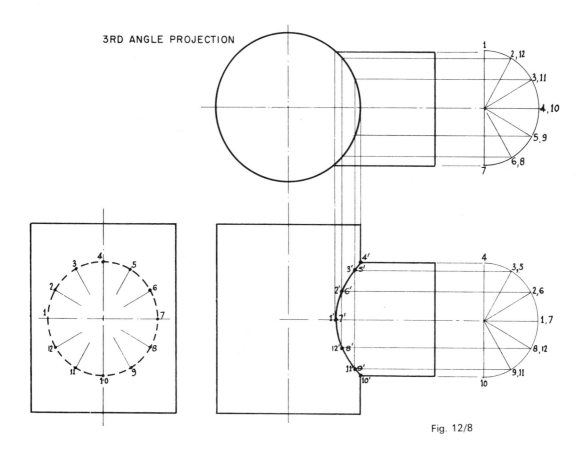

3RD ANGLE PROJECTION

Fig. 12/8

117

Two dissimilar cylinders meeting at an angle (Fig. 12/9)

The method is identical with that of the last problem. The smaller cylinder is divided into 12 equal sectors on the F.E. and on the plan.

The plan shows where these sectors meet the larger cylinder and these intersections are projected up to the F.E. to meet their corresponding sectors at 1′, 2′, 3′, etc.

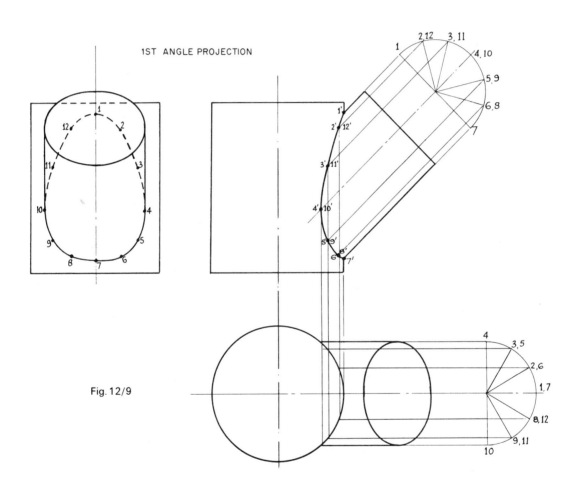

1ST ANGLE PROJECTION

Fig. 12/9

Two dissimilar cylinders meeting at an angle, their centres not being in the same vertical plane (Fig. 12/10)

Once again, the method is identical with that of the previous example. The smaller cylinder is divided into 12 equal sectors on the F.E. and on the plan.

The plan shows where the sectors meet the larger cylinder and these intersections are projected down to the F.E. to meet their corresponding sectors at 1', 2', 3', etc.

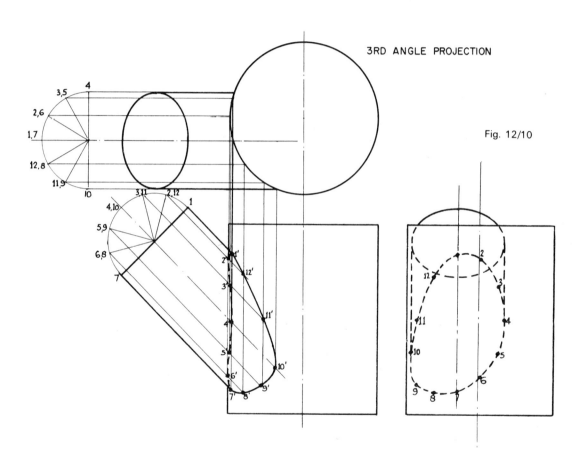

3RD ANGLE PROJECTION

Fig. 12/10

A cylinder meeting a square pyramid at right angles
(Fig. 12/11)

The F.E. shows where points 1 and 7 meet the pyramid and these are projected down to the plan.

Consider the position of point 2. Since the cylinder and the pyramid interpenetrate, point 2 lies on both the cylinder and the pyramid. Its position on the cylinder is seen fairly easily. On the F.E. it lies on the line marked 2,12 and on the plan it lies on the line marked 2,6. Its position on the pyramid is not quite so obvious. Imagine, on the F.E. that the part of the pyramid that is above the line 2,12 was removed. The section that resulted across the pyramid would be a square and point 2 would lie somewhere along the perimeter of that square. It isn't necessary to construct a complete, shaded section across the pyramid at line 2,12 but the square that would result from such a section is constructed on the plan. In Fig. 12/11 this is marked as 'SQ 2,12'. Since point 2 lies somewhere along the line 2,6 (in the plan) then its exact position is at the intersection of the square and the line. This is shown on the plan as 2'.

Point 12' is the intersection of the same square and the line 8,12 (in the plan).

This process is repeated for each point in turn. When the intersection has been completed in the plan, it is a simple matter to project the points up on to the F.E. and draw the intersection there.

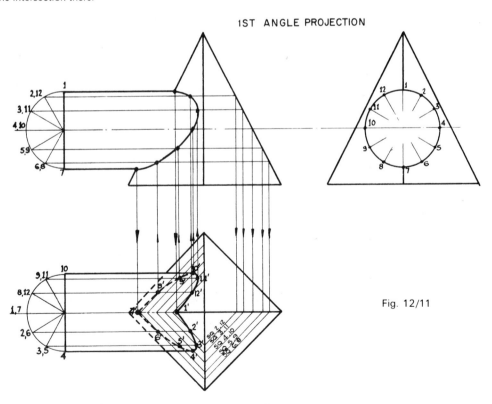

Fig. 12/11

120

A cylinder meeting a square pyramid at an angle
(Fig. 12/12)
The F.E. shows where points 1 and 7 meet the pyramid and these are projected down to the plan.

Consider the position of the point 2. In the F.E. it lies somewhere along the line marked 2,12 whilst in the plan it lies on the line marked 6,2. If that part of the pyramid above the line, 2,12 in the F.E. was removed, the point 2 would lie on the perimeter of the resulting section. This perimeter can be drawn on the plan and, in Fig. 12/12, it is shown as the line marked 'SECT 2,12'. Point 2 must lie on this line; it must also lie on the line marked 6,2 and its exact position is the intersection of these two lines.

Point 12' is the intersection of the same section line and the line marked 8,12.

This process is repeated for each point in turn. When the plan is complete the intersection can be projected onto the other two views. For the sake of clarity, these projections are not shown.

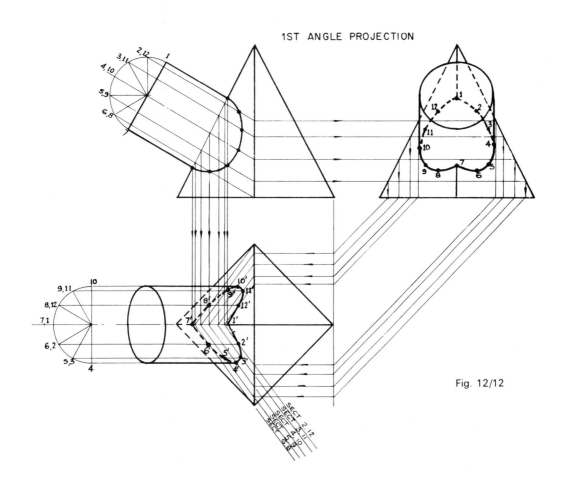

1ST ANGLE PROJECTION

Fig. 12/12

121

A cylinder meeting a hexagonal pyramid at an angle (Fig. 12/13)

Once again lines are drawn on the plan which represent the perimeters of sections taken on the F.E. on lines 1; 2, 12; 3, 11; etc. *All* the construction lines on Fig. 12/13 are for finding these section perimeters.

The line of interpenetration, first drawn on the plan, is the intersection of the line 1,7 with section 1, line 2,6 with section 2,12 (giving point 2'), line 3,5 with section 3,11 (giving point 3'), line 4 with section 4,10, etc.

When the intersection is complete on the plan, it can be projected onto the other two elevations. For the sake of clarity, these projections are not shown.

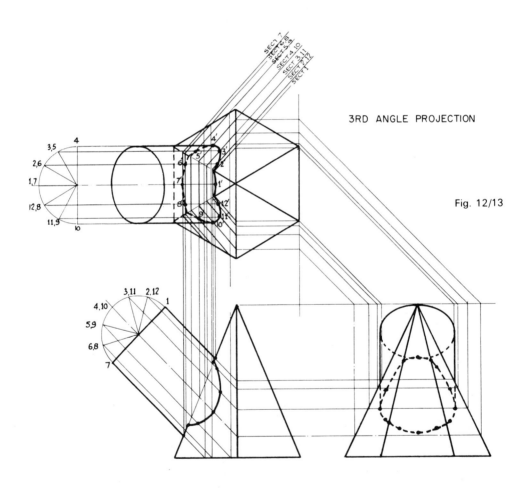

3RD ANGLE PROJECTION

Fig. 12/13

122

A cylinder meeting a cone, the cone enveloping the cylinder (Fig. 12/14)

The cylinder is divided into 12 equal sectors on the F.E. and on the plan.

Consider point 2. On the F.E. it lies somewhere along the line marked 2,12 whilst on the plan it lies on the line marked 2,6. If, on the F.E., that part of the cone above the line 2,12 was removed, point 2 would lie somewhere on the perimeter of the resulting section. In this case, the section of the cone is a circle and the radius of that circle is easily projected up to the plan. In Fig. 12/14, the section is marked on the plan as 'SECT 2,12' and the exact position of point 2' is the intersection of that section and the line marked 2,6. Point 12' is the intersection of the same section and the line marked 12,8.

This process is repeated for each point in turn. When the plan is complete, the points can be projected down to the F.E.; this is not shown for clarity.

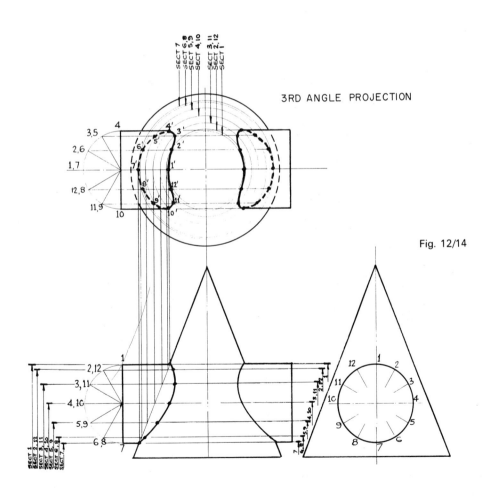

3RD ANGLE PROJECTION

Fig. 12/14

A cylinder and a cone, neither enveloping the other
(Fig. 12/15)

The constructions are exactly the same as those used in the previous example with one small addition.

The E.E. shows a point of tangency between the cylinder and the cone. This point is projected across to the F.E. and up to the plan as shown.

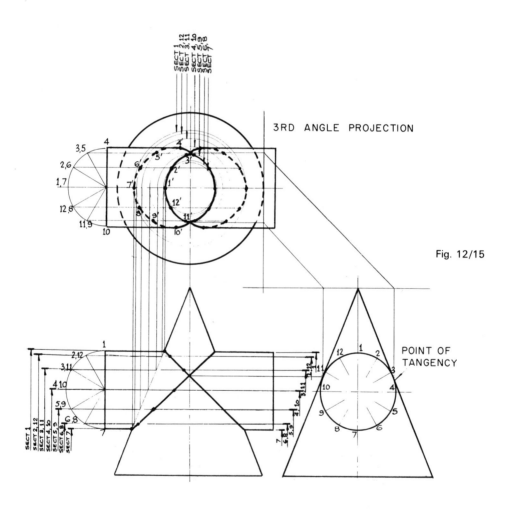

3RD ANGLE PROJECTION

Fig. 12/15

POINT OF TANGENCY

A cylinder and a cone, the cylinder enveloping the cone (Fig. 12/16)

The construction required here is a modified version of the two previous ones. Instead of the cylinder being divided into 12 equal sectors, some of which would not be used, a number of points are selected on the E.E. These are marked on the top part of the cylinder as *a, b* and *c* whilst the lower part is marked 1, 2, 3 and 4.

As before the sections of the cone across each of these points are projected up to the plan from the F.E. Each point is then projected from the E.E. to meet its corresponding section on the plan at *a', b', c', 1', 2', 3'* and *4'*.

These points are then projected down to the F.E. For the sake of clarity, this is not shown.

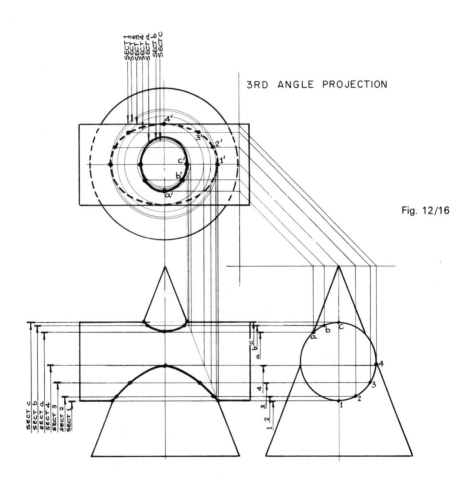

3RD ANGLE PROJECTION

Fig. 12/16

125

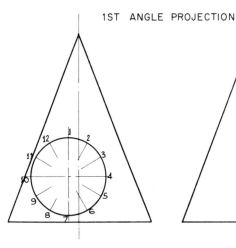

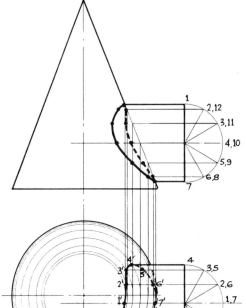

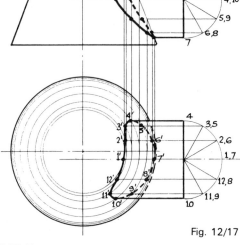

A cylinder meeting a cone, their centres not being in the same vertical plane (Fig. 12/17)

Divide the cylinder into 12 equal sectors on the F.E. and on the plan.

Sections are projected from the F.E. to the plan, the section planes being level with lines 1; 2, 12; 3, 11; 4, 10, etc. These sections appear on the plan as circles. On the plan the sectors from the cylinder are projected across to meet their respective section at points 1', 2', 3', etc. The complete interpenetration can then be projected up to the F.E. For the sake of clarity, this construction is not shown.

Fig. 12/17

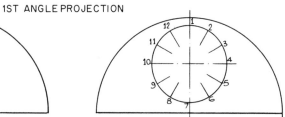

Fig. 12/18

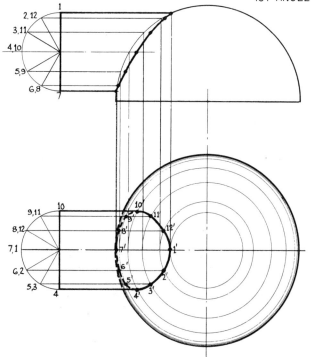

A cylinder meeting a hemisphere (Fig. 12/18)

The cylinder is divided into 12 equal sectors on the F.E. and on the plan.

Sections are projected onto the plan from the F.E. The section planes are level with the lines 1; 2,12; 3,11; 4,10, etc. and these sections appear on the plan as circles.

On the plan, the sectors from the cylinders are projected across to meet their respective section at 1', 2', 3', 4' etc.

When the interpenetration is complete on the plan, it can be projected up to the F.E. For the sake of clarity this construction is not shown.

126

A cylinder meeting a hemisphere (Fig. 12/19)
The solution is exactly the same as the last example except the sections are projected onto the E.E. and not the plan.

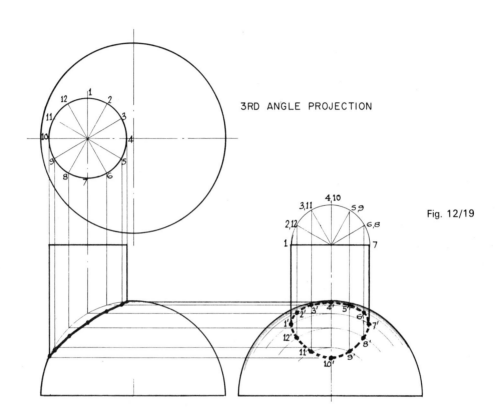

3RD ANGLE PROJECTION

Fig. 12/19

FILLET CURVES
A sudden change of shape in any load-bearing component produces a stress centre, that is, an area that is more highly stressed than the rest of the component and therefore more liable to fracture under load. To avoid these sharp corners, fillet radii are used. These radii allow the stress to be distributed more evenly, making the component stronger.

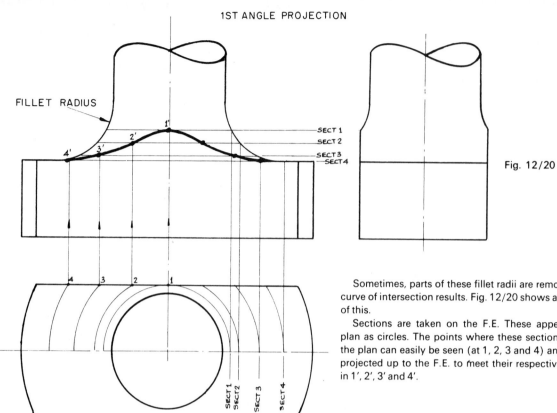

FILLET RADIUS

SECT 1
SECT 2
SECT 3
SECT 4

Fig. 12/20

Sometimes, parts of these fillet radii are removed and a curve of intersection results. Fig. 12/20 shows an example of this.

Sections are taken on the F.E. These appear on the plan as circles. The points where these sections 'run off' the plan can easily be seen (at 1, 2, 3 and 4) and they are projected up to the F.E. to meet their respective sections in 1', 2', 3' and 4'.

3RD ANGLE PROJECTION

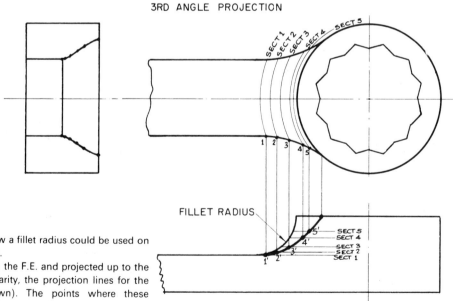

FILLET RADIUS

SECT 5
SECT 4
SECT 3
SECT 2
SECT 1

Fig. 12/21 shows how a fillet radius could be used on the end of a ring spanner.

Sections are taken on the F.E. and projected up to the plan (for the sake of clarity, the projection lines for the sections are not shown). The points where these sections 'run off' the plan can easily be seen and these points (1, 2, 3, 4 etc.) are projected down to the F.E. to meet their respective sections in 1', 2', 3', 4', etc.

Fig. 12/21

Exercises 12

(All questions originally set in Imperial units)

1. Fig. 1 shows the plan and incomplete elevation of two cylinders. Draw the two views, showing any hidden lines.

North Western Secondary School Examinations Board

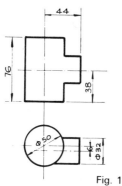

Fig. 1

DIMENSIONS IN mm

2. The plan and incomplete elevation of two pipes are shown in Fig. 2. Copy the two views, full size, and complete the elevation showing hidden detail.

Middlesex Regional Examining Board

Fig. 2

3. Fig. 3 shows an incomplete elevation of the junction of a cylinder and an equilateral triangular prism. The axes of both lie in the same vertical plane. The prism rests with one of its side faces in the H.P.

Draw, and complete, the given elevation and project a plan. Do *not* show hidden detail.

Joint Matriculation Board

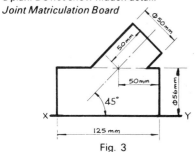

Fig. 3

4. Fig. 4 shows incomplete drawings of the plan and elevation of a junction between a square section pipe and a cylindrical pipe.

Draw (a) the complete plan and elevation and (b) the development of the whole surface of *either* the square pipe *or* the cylindrical one.

Southern Universities' Joint Board (See Ch. 14 for information not in Ch. 12)

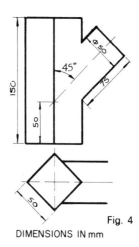

Fig. 4

DIMENSIONS IN mm

5. Fig. 5 gives the plan and incomplete elevation of a junction between a cylinder and a square prism. Copy the two views and complete the elevation showing all hidden detail.

Oxford and Cambridge Schools Examination Board

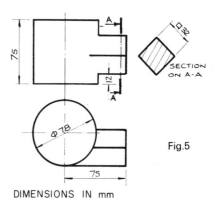

Fig.5

DIMENSIONS IN mm

129

6. Fig. 6 consists of a plan and incomplete elevation of a square prism intersecting a cone.
   (a) Draw the given plan.
   (b) Draw a complete elevation showing the curve of intersection.
   (c) Develop the surface of the cone below the curve of intersection.
   Southern Universities' Joint Board (See Ch. 14 for information not in Ch. 12)

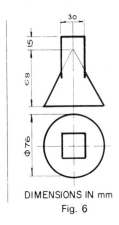

DIMENSIONS IN mm

Fig. 6

7. Two elevations of a wheel stop are shown in Fig. 7. Draw, full size, (a) the given elevations, (b) a plan looking in the direction of arrow Z.
   Associated Examining Board

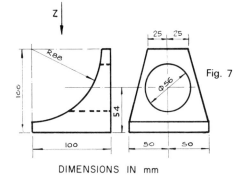

Fig. 7

DIMENSIONS IN mm

8. Fig. 8 shows an incomplete plan view of the junction of a cylinder to a right square pyramid. The axis of the cylinder is 32 mm above the base of the pyramid which stands on the H.P. Perpendicular height of the pyramid = 100 mm.
   Draw and complete the given plan and project an elevation looking in the direction of arrow A.
   Draw, also, the development of the pyramid portion showing the hole required to receive the cylinder. Show all hidden detail.
   Joint Matriculation Board (See Ch. 14 for information not in Ch. 12)

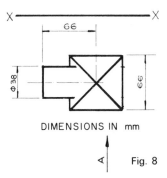

DIMENSIONS IN mm

Fig. 8

9. The height of a right circular cone is 88 mm and the base diameter is 94 mm. The cone is pierced by a square hole of side 32 mm. The axis of the hole intersects the axis of the cone 32 mm above the base, and is parallel to the base.
   Draw an elevation of the cone looking in a direction at right angles to the vertical faces of the hole.
   Oxford and Cambridge Schools Examination Board

10. The plan and incomplete elevation of a solid are shown in Fig. 9. Reproduce the given views and complete the elevation by including the lines of intersection produced by the vertical faces A and B. Hidden edges are to be shown.
    Cambridge Local Examinations

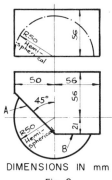

DIMENSIONS IN mm

Fig. 9

11. Two views (one incomplete) of a connecting rod end are shown in Fig. 10. The original diameters were 126 mm and 62 mm and the transition between these followed a circular path of radius 75 mm. Two flat parallel faces were then milled as shown in the end elevation. Draw the given views, complete the left-hand elevation and beneath this project a plan. Scale—full size.

Oxford Local Examinations

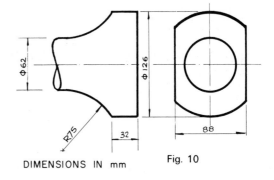

DIMENSIONS IN mm

Fig. 10

13

Further orthographic projection

The first five diagrams of Chapter 10 show the basic principles of orthographic projection. These diagrams should be thoroughly understood before this chapter is attempted. The object to be drawn is suspended between three planes called the front vertical plane (F.V.P.), the end vertical plane (E.V.P.) and the horizontal plane (H.P.). These planes are at right angles, and a view is projected on to each of the planes. These views are called the front elevation (F.E.), the end elevation (E.E.), and the plan. Two of the planes are then swung back, as if on hinges, until all three planes are in the same plane, i.e. they would all lie on the same flat surface. This system of swinging the planes until they are in line is called *rabatment*.

Definitions

When a line passes through a plane, the point of intersection is called a 'trace'.

When a plane passes through another plane, the line of intersection is also called a trace.

THE STRAIGHT LINE

The projection of a line which is not parallel to any of the principal planes

Fig. 13/1 shows a straight line AB suspended between the three principal planes. Projectors from A and B, perpendicular to the planes, give the projection of AB on each of the principal planes. On the right of Fig. 13/1 can be seen the projections of the line after rabatment.

A trace is the line resulting from the intersection of two planes. The trace of the F.V.P. and the H.P. is the line OX. The trace of the E.V.P. and the H.P. is the line OY. The trace of the F.V.P. and the E.V.P. is the line OZ. These lines are often very useful for reference purposes and they should be marked on your drawings. The O is often ignored and the traces are then shown as XY and YZ.

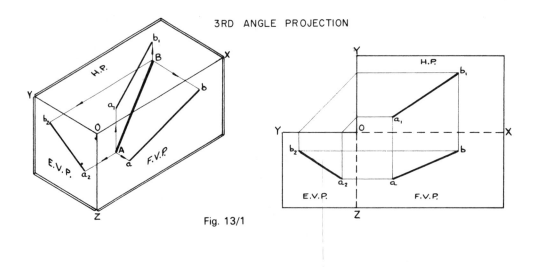

3RD ANGLE PROJECTION

Fig. 13/1

To find the true length of a line that is not parallel to any of the principal planes and to find the angle that the line makes with the F.V.P. (Fig. 13/2)

The line is AB. On the F.V.P. it is seen as *ab* and on the H.P. as a_1b_1.

One end of the line A is kept stationary whilst B is swung round so that AB is parallel to the H.P. B is now at B' and on the F.V.P. *b* is now at *b'*. Since the line is parallel to the H.P. it will project its true length onto the H.P. This is shown as a_1b_2. Notice that b_1 and b_2 are the same distance from the line XY.

Since AB' (and *ab'*) are parallel to the H.P., the angle that AB makes with the F.V.P. can be measured. This is shown as θ.

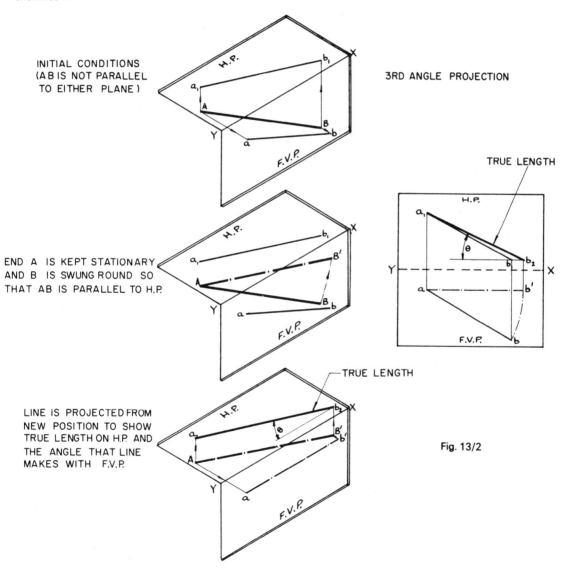

INITIAL CONDITIONS
(AB IS NOT PARALLEL
TO EITHER PLANE)

3RD ANGLE PROJECTION

TRUE LENGTH

END A IS KEPT STATIONARY
AND B IS SWUNG ROUND SO
THAT AB IS PARALLEL TO H.P.

TRUE LENGTH

LINE IS PROJECTED FROM
NEW POSITION TO SHOW
TRUE LENGTH ON H.P. AND
THE ANGLE THAT LINE
MAKES WITH F.V.P.

Fig. 13/2

133

To find the true length of a line that is not parallel to any of the principal planes and to find the angle that the line makes with the H.P. (Fig. 13/3)

The line is AB. On the F.V.P. it is seen as *ab* and on the H.P. as a_1b_1.

One end of the line B is kept stationary whilst A is swung round so that AB is parallel to the F.V.P. A is now at A' and, on the H.P., *a* is now at *a'*. Since the line is now parallel to the F.V.P. it will project its true length onto the F.V.P. This is shown as a_2b. Notice that a_2 and *a* are the same distance from the line XY.

Since BA' (and b_1a') are parallel to the F.V.P., the angle that AB makes with the H.P. can be measured. This is shown as ϕ.

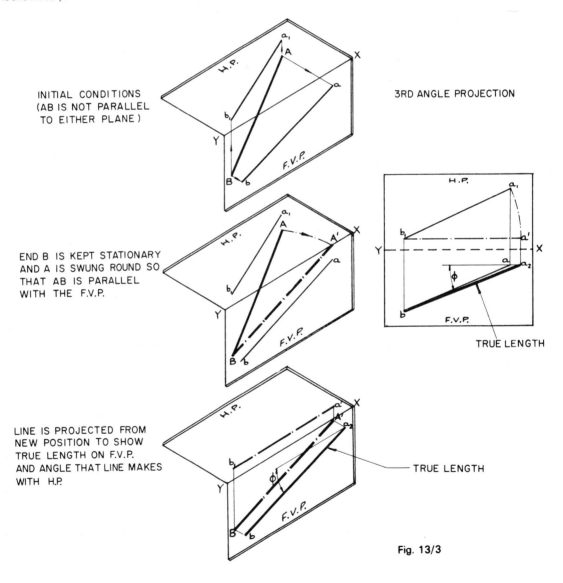

INITIAL CONDITIONS
(AB IS NOT PARALLEL
TO EITHER PLANE)

3RD ANGLE PROJECTION

END B IS KEPT STATIONARY
AND A IS SWUNG ROUND SO
THAT AB IS PARALLEL
WITH THE F.V.P.

TRUE LENGTH

LINE IS PROJECTED FROM
NEW POSITION TO SHOW
TRUE LENGTH ON F.V.P.
AND ANGLE THAT LINE MAKES
WITH H.P.

TRUE LENGTH

Fig. 13/3

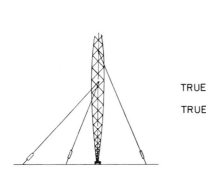

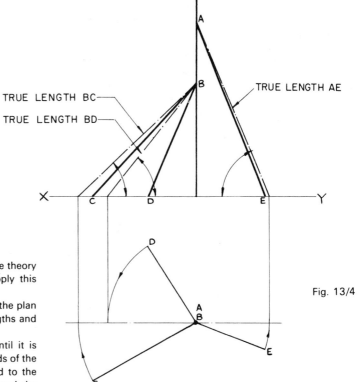

TRUE LENGTH BC

TRUE LENGTH BD

TRUE LENGTH AE

Fig. 13/4

Fig. 13/4 is an example of an application of the theory shown above. It shows how simple it is to apply this theory.

A pylon is supported by three hawsers. Given the plan and elevation of the hawsers, find their true lengths and the angle that they make with the ground.

In the plan, each hawser is swung round until it is parallel to the F.V.P. The new positions of the ends of the hawsers are projected up to the F.E. and joined to the pylon at A and B. This gives the true lengths and the angles.

To find the traces of a straight line given the plan and elevation of the line (Fig. 13/5)

The line is AB. If the line is produced it will pass through both planes, giving traces T_v and T_h.

ab is produced to meet the XY line. This intersection is projected down to meet a_1b_1 produced in T_h.

b_1a_1 is produced to meet the XY line. This intersection is projected up to meet ba produced in T_v.

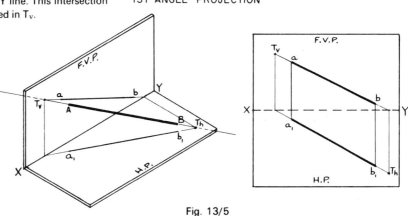

Fig. 13/5

To draw the elevation and plan of a line AB given its true length and the distances of the ends of the line from the principal planes, in this case a_v and a_h, and b_v and b_h (Fig. 13/6)
1. Fix points a and a_1 at the given distances a_v and a_h from the XY line. These are measured on a common perpendicular to XY.
2. Draw a line parallel to XY, distance b_v from XY.
3. With centre a, radius equal to the true length AB, draw an arc to cut the line drawn parallel to XY in C.
4. From a_1 draw a line parallel to XY to meet a line from C drawn perpendicular to XY in D.
5. Draw a line parallel to XY, distance b_h from XY.
6. With centre a_1, radius a_1D, draw an arc to cut the line drawn parallel to XY in b_1.
7. Draw a line from b_1, perpendicular to XY to meet the line drawn parallel to XY through C in b.
ab is the elevation of the line.
a_1b_1 is the plan of the line.

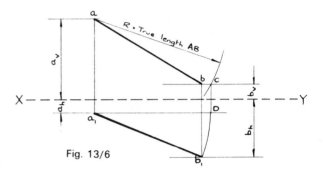

1ST ANGLE PROJECTION

Fig. 13/6

To construct the plan of a line AB given the distance of one end of the line from the XY line in the plan (a_h), the true length of the line and the elevation (Fig. 13/7)
1. From b draw a line parallel to the XY line.
2. With centre a, radius equal to the true length of the line AB, draw an arc to cut the parallel line in C.
3. From a_1 (given), draw a line parallel to the XY line to meet a line drawn from C perpendicular to XY in D.
4. With centre a_1, radius equal to a_1D draw an arc to meet a line drawn from b perpendicular to XY in b_1.
a_1b_1 is the plan of the line.

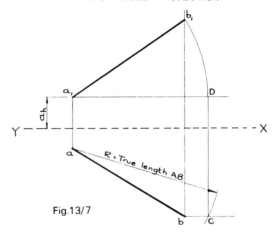

3RD ANGLE PROJECTION

Fig.13/7

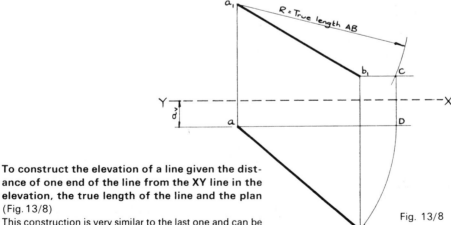

3RD ANGLE PROJECTION

Fig. 13/8

To construct the elevation of a line given the distance of one end of the line from the XY line in the elevation, the true length of the line and the plan (Fig. 13/8)
This construction is very similar to the last one and can be followed from the instructions given for that example.

To construct the elevation of a line AB given the plan of the line and the angle that the line makes with the horizontal plane (Fig. 13/9)

1. Draw the plan and from one end erect a perpendicular.
2. From the other end of the plan draw a line at the angle given to intersect the perpendicular in C.
3. From b_1 draw a line perpendicular to XY to meet XY in b.
4. From a_1 draw a line perpendicular to XY and mark off XY to a equal to a_1c.

ab is the required elevation. An alternative solution is also shown.

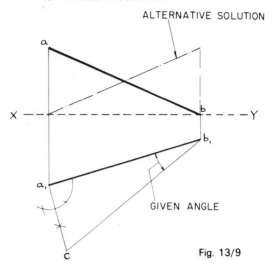

Fig. 13/9

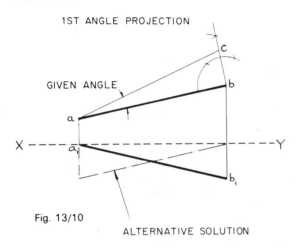

Fig. 13/10

To draw the plan of a line AB given the elevation of the line and the angle that the line makes with the vertical plane (Fig. 13/10)

This construction is very similar to the last one and can be followed from the instructions given for that example.

ALTERNATIVE SOLUTION

3RD ANGLE PROJECTION

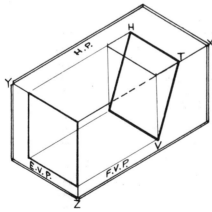

THE INCLINED PLANE

Definition

An inclined plane is inclined to two of the principal planes and perpendicular to the third.

Fig. 13/11 shows a rectangular plane that is inclined to the H.P. and the E.V.P. and is perpendicular to the F.V.P. Because it is perpendicular to the F.V.P., the true angle between the inclined plane and the H.P. can be measured on the F.V.P. This is the angle ϕ.

TRACES OF PLANE

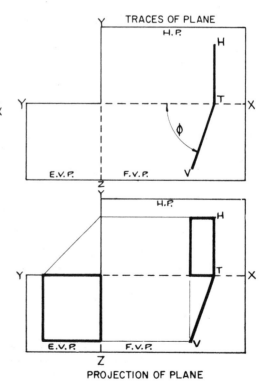

PROJECTION OF PLANE

Fig. 13/11

The top drawing shows the traces of the plane after rabatment. The bottom drawing shows the full projection of the plane. It should be obvious how the full projection is obtained if you are given the traces and told that the plane is rectangular.

Fig. 13/12 shows a triangular plane inclined to the F.V.P. and the E.V.P., and perpendicular to the H.P. Because it is perpendicular to the H.P., the true angle between the inclined plane and the F.V.P. can be measured on the H.P. This angle is θ.

Once again, it should be obvious how the full projection of the inclined plane is obtained if you are given the traces and told that the plane is triangular.

3RD ANGLE PROJECTION

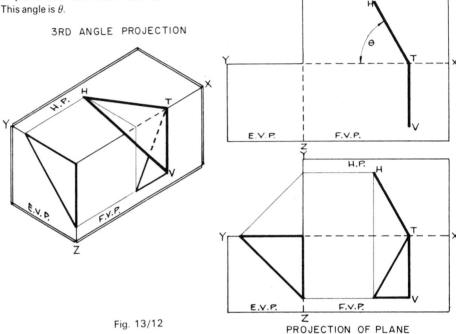

Fig. 13/12

TRACES OF PLANE

PROJECTION OF PLANE

To find the true shape of an inclined plane
If the inclined plane is swung round so that it is parallel to one of the reference planes, the true shape can be projected. In Fig. 13/13, the plan of the plane, HT, is swung round to H'T. The true shape of the plane can then be drawn on the F.V.P.

3RD ANGLE PROJECTION

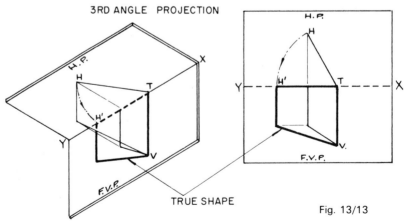

TRUE SHAPE

Fig. 13/13

Fig. 13/14 shows an example. An oblique, truncated, rectangular pyramid stands on its base. The problem is to find the true shape of sides A and B.

In the F.E. side A is swung upright and its vertical height is projected across to the E.E. where the true shape of side A can be drawn.

In the E.E. side B is swung upright and projected across to the F.E. where the true shape is drawn.

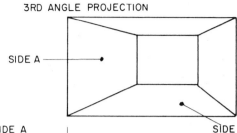

3RD ANGLE PROJECTION

SIDE A

SIDE B

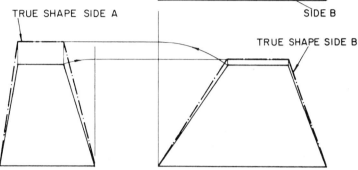

TRUE SHAPE SIDE A

TRUE SHAPE SIDE B

Fig. 13/14

THE OBLIQUE PLANE

Definition

An oblique plane is a plane that is inclined to all of the principal planes.

Fig. 13/15 shows a quadrilateral plane that is inclined to all three principal planes.

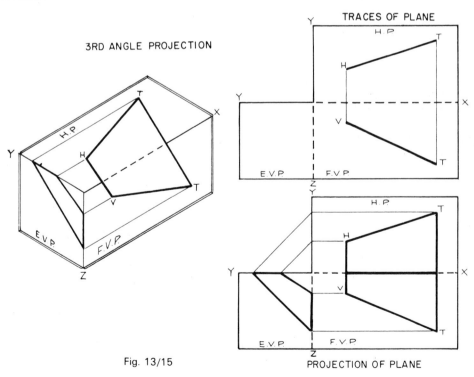

3RD ANGLE PROJECTION

TRACES OF PLANE

PROJECTION OF PLANE

Fig. 13/15

The top drawing shows the traces of the plane after rabatment. The bottom drawing shows the full projection of the plane. It should be obvious how the projection is obtained if you are given the traces.

Fig. 13/16 shows a triangular plane that is inclined to all three principal planes.

The top drawing shows the traces of the plane after rabatment. The bottom drawing shows the full projection of the plane. It should be obvious how the projection is obtained if you are given the traces.

3RD ANGLE PROJECTION

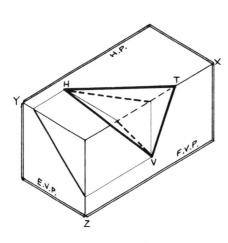

Fig. 13/16

TRACES OF PLANE

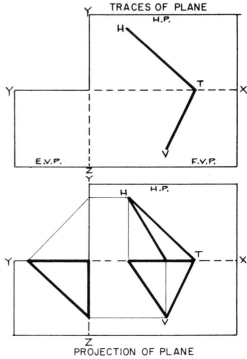

PROJECTION OF PLANE

To find the true angle between the H.P. and an oblique plane (Fig. 13/17)

A triangle is inserted under the oblique plane and at right angles to it. This triangle, PQR, meets the H.P. at the same angle as the oblique plane.

The triangle is swung round until it is parallel to the F.V.P. Its new position is PQS and the angle required is PŜQ.

1ST ANGLE PROJECTION

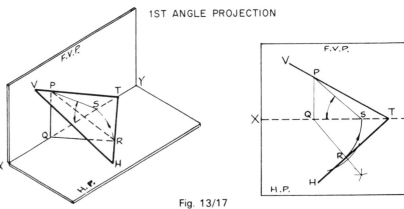

Fig. 13/17

140

To find the true angle between the V.P. and an oblique plane (Fig. 13/18)

A triangle is inserted under the oblique plane and at right angles to it. This triangle, PQR, meets the F.V.P. at the same angle as the oblique plane.

The triangle is swung round until it is parallel to the H.P. Its new position is PQS and the angle required is PŜQ.

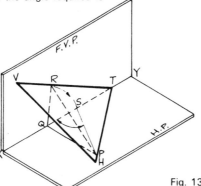

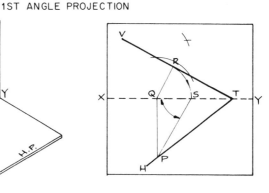

Fig. 13/18

To find the true angle between the traces of a given oblique plane VTH (Fig. 13/19)

From any point a on the XY line draw ab (perpendicular to XY) and ac (perpendicular to TH).

With centre T, radius Tb, draw an arc to meet ac produced in d.

dT̂c is the required angle.

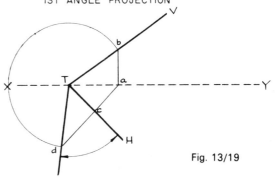

Fig. 13/19

Exercises 13

(All questions originally set in Imperial units)

1. Fig. 1 shows the shaped end of a square fence post. Lengths AD, BD, DC are equal.

   By means of a suitable geometric construction find (a) the true length of BD, (b) the true shape of the face ABD.

   Associated Lancashire Schools Examining Board

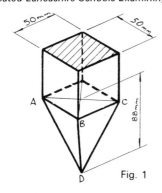

Fig. 1

2. Fig. 2 gives the front elevation and plan of the roof of a house in 1st angle projection. Draw the given views to a suitable scale and from them find, by construction, the true lengths of the rafters A and B. Print these lengths under your drawing.

   East Anglian Regional Examinations Board

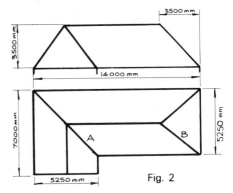

Fig. 2

141

3. Fig. 3 shows the elevation of a line AB which has a true length of 100 mm. End B of the line is 12 mm in front of the V.P.; end A is also in front of the V.P. Draw the plan and elevation of this line and determine and indicate its V.T. and H.T. Measure, state and indicate the angle of inclination of the line to the H.P. *Joint Matriculation Board*

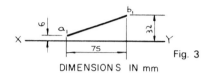

Fig. 3

DIMENSIONS IN mm

4. A line AB of true length 88 mm lies in an auxiliary vertical plane which makes an angle of 30° with the vertical plane. The line is inclined at an angle of 45° to the horizontal, the point B being the lowest at a vertical distance of 12 mm above the horizontal plane and 12 mm in front of the vertical plane. Draw the plan and elevation of AB and clearly identify them in the drawing. *Associated Examining Board*

5. The plan of a line 82 mm long is shown in Fig. 4. The elevation of one end is at b'. Complete the elevation and measure the inclinations of the line to the H.P. and V.P. *University of London School Examinations*

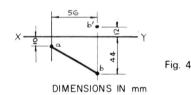

Fig. 4

DIMENSIONS IN mm

6. Fig. 5 shows the plan of end A of a line AB. End A is in the plane H.T.V. End B is in the H.P. The line AB is perpendicular to the plane H.T.V. Draw the plan and elevation of AB. *Joint Matriculation Board* (H.V.T. = Horizontal and vertical traces)

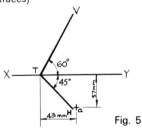

Fig. 5

7. The projections of a triangle RST are shown in Fig. 6. Determine the true shape of the triangle. *Associated Examining Board*

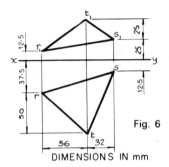

Fig. 6

DIMENSIONS IN mm

8. The plan and elevation of two straight lines are given in Fig. 7. Find the true lengths of the lines, the true angle between them and the distance between A and C. *Southern Universities' Joint Board*

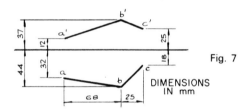

Fig. 7

DIMENSIONS IN mm

9. Fig. 8 shows the plan and elevation of a triangular lamina. Draw these two views and, by finding the true length of each side, draw the true shape of the lamina. Measure and state the three angles to the nearest degree. *Oxford and Cambridge Schools Examination Board*

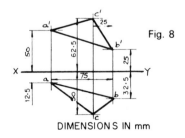

Fig. 8

DIMENSIONS IN mm

142

10. At Fig. 9 are shown the elevation and plan of a triangle ABC. Determine the true shape and size of the triangle.
University of London School Examinations

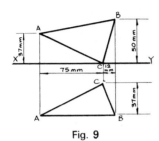

Fig. 9

11. Fig. 10 shows two views of an oblique triangular pyramid, standing on its base. Draw the given view together with an auxiliary view looking in the direction of arrow A which is perpendicular to BC. Also draw the true shapes of the sides of the pyramid. Note: Omit dimensions but show hidden detail in all views. Scale: full size.
Oxford Local Examinations (See Ch. 10 for information not in Ch. 13)

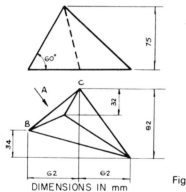

DIMENSIONS IN mm

Fig. 10

14
Developments

There are two basic ways of fashioning a piece of material into a given shape. Either you start with a solid lump and take pieces off until the required shape is obtained or you have the material in sheet form and bend it to the required shape.

It should be obvious that, if the latter method is used, the sheet material must first be shaped so that, after it is bent, you have the correct size and shape. If, then a component is to be made of sheet material, the designer must not only visualize and draw the final three-dimensional component, he must also calculate and draw the shape of the component in the form that it will take when marked out on the two-dimensional sheet material.

The process of unfolding the three-dimensional 'solid' is called *development*.

The shapes of most engineering components are whole, or parts of, prisms, pyramids, cylinders or cones and so this chapter deals with the development of the shapes.

PRISMS

Fig. 14/1 shows how a square prism is unfolded and its development obtained.

Notice that where there are corners in the undeveloped solid, these are shown as dotted lines in the development.

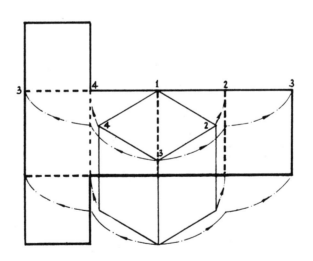

Fig. 14/1

To develop a square prism with an oblique top

This development, shown in Fig. 14/2 should be self-explanatory.

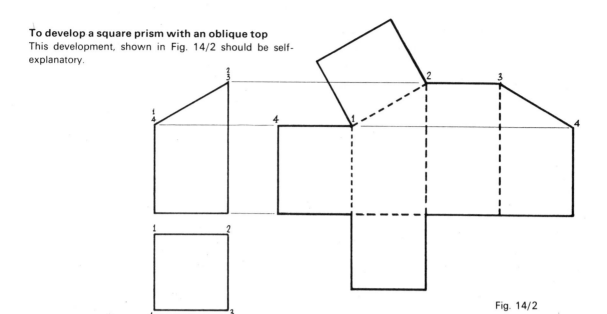

Fig. 14/2

To develop a hexagonal prism with oblique ends (Fig. 14/3)

The height of each corner of the development is found by projecting directly from the orthographic view.

The shapes of the top and the bottom are found by projecting the true shapes of the oblique faces. The top has been found by conventional means. The true shape is projected from the elevation and transferred to the development.

The true shape of the bottom of the prism has been drawn directly on the development without projecting the true shape from the elevation. The corner between lines 2 and 3 has been produced until it meets the projectors from corners 1 and 4. The produced line is then turned through 90° and the width, 2A, marked on.

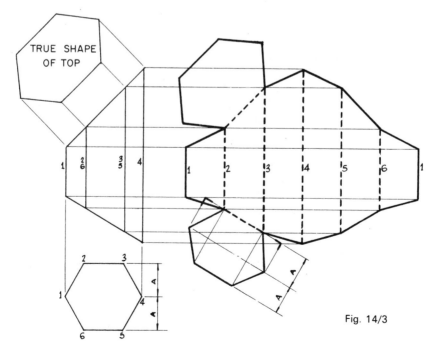

TRUE SHAPE OF TOP

Fig. 14/3

The development of intersecting square and hexagonal prisms meeting at right angles (Fig. 14/4)
First an orthographic drawing is made and the line of interpenetration is plotted. The development of the hexagonal prism is projected directly from the F.E. and the development of the square prism is projected directly from the plan.

Projecting from the orthographic views provides much of the information required to develop the prisms; any other information can be found on one of the orthographic views and transferred to the developments. In this case, dimensions A, b, C and d have not been projected but have been transferred with dividers.

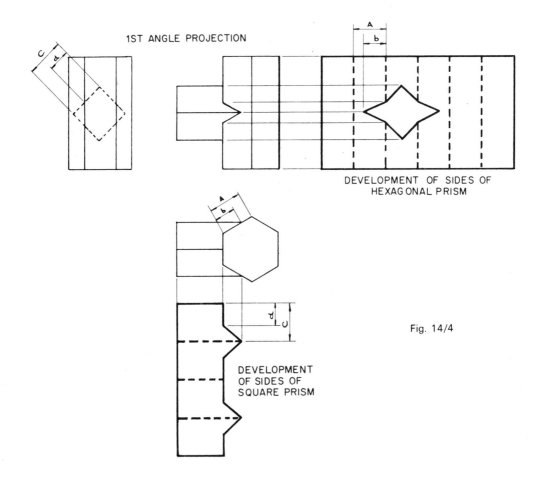

1ST ANGLE PROJECTION

DEVELOPMENT OF SIDES OF HEXAGONAL PRISM

DEVELOPMENT OF SIDES OF SQUARE PRISM

Fig. 14/4

146

The development of intersecting hexagonal and octagonal prisms meeting at an angle (Fig. 14/5)
The method of developing these prisms is identical to that used in the previous example. This example is more complicated but the developments are still projected from one of the orthographic views, and any information which is not projected across can be found on the orthographic views and transferred to the development. In this case, dimensions A, *b, c,* D, etc. have not been projected but have been transferred with dividers.

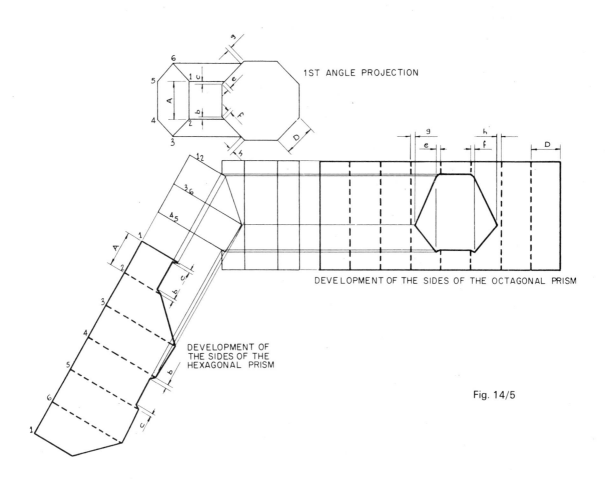

1ST ANGLE PROJECTION

DEVELOPMENT OF THE SIDES OF THE OCTAGONAL PRISM

DEVELOPMENT OF THE SIDES OF THE HEXAGONAL PRISM

Fig. 14/5

147

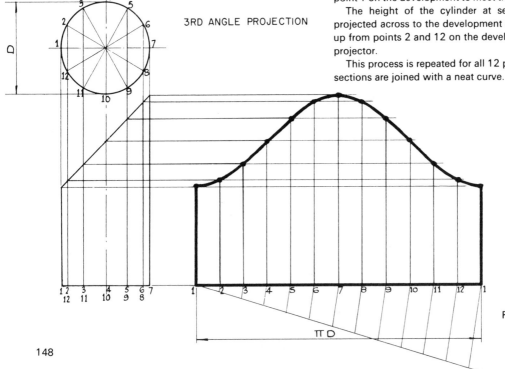

CYLINDERS

If you painted the curved surface of a cylinder and, whilst the paint was wet, placed the cylinder on a flat surface and then rolled it once, the pattern that the paint left on the flat surface would be the development of the curved surface of the cylinder. Fig. 14/6 shows the shape that would evolve if the cylinder was cut obliquely at one end. The length of the development would be ΠD, the circumference.

The oblique face has been divided into twelve equal parts and numbered. You can see where each number will touch the flat surface as the cylinder is rolled.

Fig. 14/6

Fig. 14/7 shows how the above idea is interpreted into an accurate development of a cylinder.

To develop a cylinder with an oblique top (Fig. 14/7)

A plan and elevation of the cylinder is drawn. The plan is divided into 12 equal sectors which are numbered. These numbers are also marked on the elevation.

The circumference of the cylinder is calculated and is marked out alongside the elevation. This circumference ΠD is divided into 12 equal parts and these parts are numbered 1 to 12 to correspond with the twelve equal sectors.

The height of the cylinder at sector 1 is projected across to the development and a line is drawn up from point 1 on the development to meet the projector.

The height of the cylinder at sectors 2 and 12 is projected across to the development and lines are drawn up from points 2 and 12 on the development to meet the projector.

This process is repeated for all 12 points and the intersections are joined with a neat curve.

3RD ANGLE PROJECTION

Fig. 14/7

148

Fig. 14/8 shows an extension of the construction shown above.

To develop a cylinder which is cut obliquely at both ends (Fig. 14/8)

The method is identical to that used for the last example. However, the following points should be noted.

The projectors from the elevation which show the 'heights' of points 1 to 12 are projected at 90° to the ℄ of the cylinder being developed.

Only half a circle is necessary to divide the cylinder into 12 equal sectors and this must be projected so that the base of the semi-circle is at 90° to the ℄ of the cylinder.

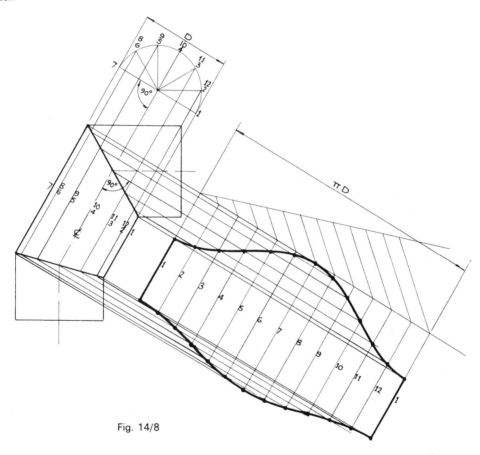

Fig. 14/8

To develop a cylinder which, in elevation, has a circular piece cut-out (Fig. 14/9)

The general method of developing a cylinder of this nature is similar to those shown above. The plan of the cylinder is divided into twelve equal sectors and the location of the sectors which are within the circular cut-out are projected down to the F.E. and across to the development.

There are some more points which must also be plotted. These are 3', 5', 9' and 11'. Their positions can be seen most easily on the F.E. and they are projected up to the plan. The plan shows how far they are away from points 3, 5, 9 and 11 and these distances, a and b, can be transferred to the development. The exact positions of these points can then be projected across from the F.E. to the development.

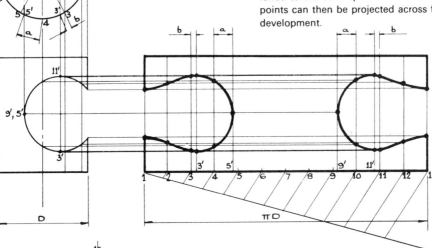

3RD ANGLE PROJECTION

Fig. 14/9

To develop an intersecting cylinder (Fig. 14/10)

The shape of the development is determined by the shape of the line of intersection. Once this has been found, the development is found using the same methods as in previous examples.

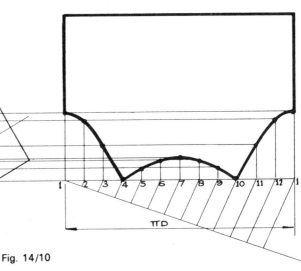

Fig. 14/10

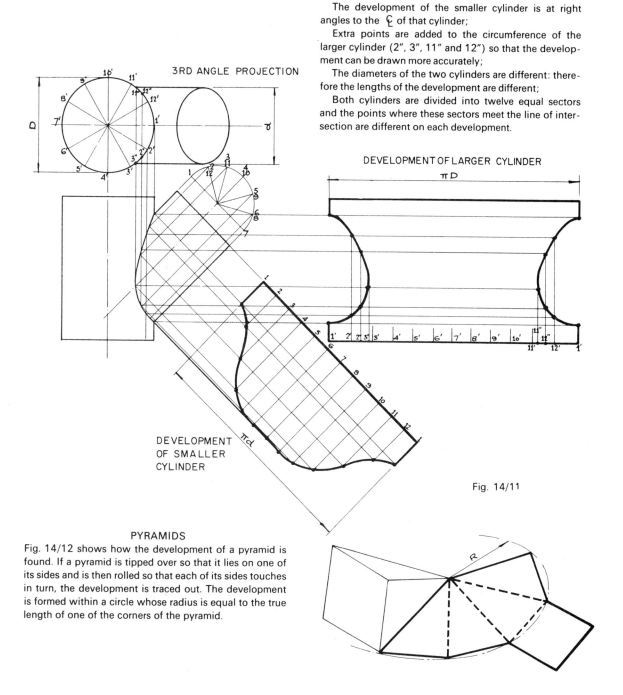

To develop both intersecting cylinders (Fig. 14/11)
The developments are found using methods discussed above. Particular points to notice are:

The development of the smaller cylinder is at right angles to the ₵ of that cylinder;

Extra points are added to the circumference of the larger cylinder (2", 3", 11" and 12") so that the development can be drawn more accurately;

The diameters of the two cylinders are different: therefore the lengths of the development are different;

Both cylinders are divided into twelve equal sectors and the points where these sectors meet the line of intersection are different on each development.

3RD ANGLE PROJECTION

DEVELOPMENT OF LARGER CYLINDER

π D

DEVELOPMENT OF SMALLER CYLINDER

π d

Fig. 14/11

PYRAMIDS

Fig. 14/12 shows how the development of a pyramid is found. If a pyramid is tipped over so that it lies on one of its sides and is then rolled so that each of its sides touches in turn, the development is traced out. The development is formed within a circle whose radius is equal to the true length of one of the corners of the pyramid.

Fig. 14/12   R=TRUE LENGTH OF A CORNER OF THE PYRAMID

To develop the sides of the frustum of a square pyramid (Fig. 14/13)

The true length of a corner of the pyramid can be seen in the F.E. An arc is drawn, radius equal to this true length, centre the apex of the pyramid. A second arc is drawn, radius equal to the distance from the apex of the cone to the beginning of the frustum, centre the apex of the cone. The width of one side of the pyramid, measured at the base, is measured on the plan and this is stepped round the larger arc four times.

1ST ANGLE PROJECTION

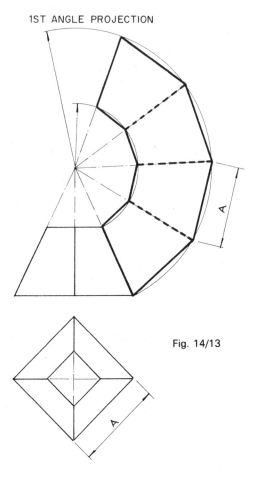

Fig. 14/13

3RD ANGLE PROJECTION

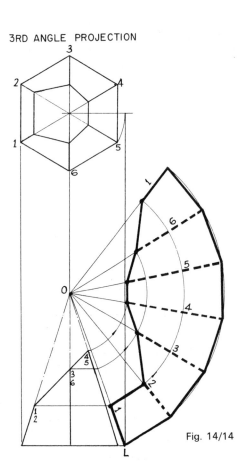

Fig. 14/14

To develop the sides of a hexagonal frustum if the top has been cut obliquely (Fig. 14/14)

The F.E. does not show the true length of a corner of the pyramid. Therefore, the true length, OL, is constructed and an arc, radius OL and centre O, is drawn. The width of one side of the pyramid, measured at the base, is stepped around the arc six times and the six sides of the pyramid are marked on the development.

The F.E. does not show the true length of a corner of the pyramid; equally it does not show the true distance from O to any of the corners 1 to 6. However, if each of these corners is projected horizontally to the line OL (the true length of a corner), these true distances will be seen. With compass centre at O, these distances are swung round to their appropriate corners.

To develop a hexagonal pyramid that has been penetrated by a square prism (Fig. 14/15)

Plot the line of interpenetration (see Chapter 12) and develop the pyramid as if it were complete.

The line of intersection between the prism and the pyramid has to be plotted on the development. Most of the changes of shape in this line occur on corners of the pyramid and it is a simple matter to plot these on the development. They are swung round with compasses from the F.E. However, there is a change of shape that does not occur on a corner. It occurs on the sides between corners 2 and 3, and 4 and 5. A line is drawn, on the plan, from the apex of the pyramid, through these corners to meet the base of the pyramid in 2a and 5a. These two points are transferred to the development and the exact positions of the corners can be plotted on the development.

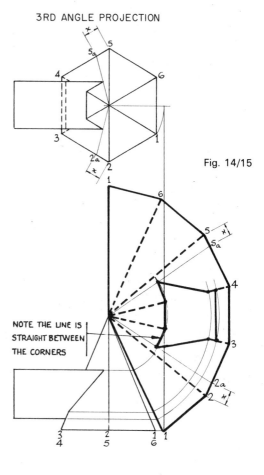

Fig. 14/15

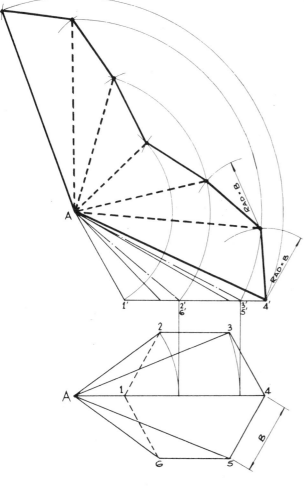

Fig. 14/16

To develop an oblique hexagonal pyramid

(An oblique pyramid, unlike a right pyramid, does not have its apex directly above the centre of its base.)

The true lengths of all the corners of the pyramid are found. In Fig. 14/16 these are shown as A1', A2', A3', etc. With centre A, radius each of the true lengths in turn, arcs are drawn.

Starting at 4', the length of one side of the base B is drawn as a radius, meeting the arc drawn from 3'5'. This intersection is used as a centre for drawing another arc of the same radius, meeting the arc drawn from 2'6'. This is repeated for the complete development.

153

CONES

Fig. 14/17 shows how, if a cone is tipped over and then rolled it will trace out its development. The development forms a sector of a circle whose radius is equal to the slant height of the cone. The length of the arc of the sector is equal to the circumference of the base of the cone.

If the base of the cone is divided into twelve equal sectors which are numbered from 1 to 12, the points where the numbers touch the flat surface as the cone is rolled can be seen.

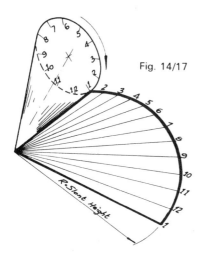

Fig. 14/17

To develop the frustum of a cone

The plan and elevation of the cone are shown in Fig. 14/18. The plan is divided into 12 equal sectors. The arc shown as dimension A is 1/12 of the circumference of the base of the cone.

With centre at the apex of the cone draw two arcs, one with a radius equal to the distance from the apex to the top of the frustum (measured along the side of the cone) and the other equal to the slant height of the cone.

With dividers measure distance A and step this dimension around the larger arc 12 times. (This will not give an exact measurement of the circumference at the base of the cone but it is a good approximation.)

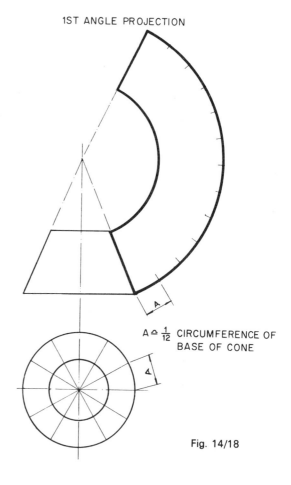

1ST ANGLE PROJECTION

$A \triangleq \frac{1}{12}$ CIRCUMFERENCE OF BASE OF CONE

Fig. 14/18

To develop the frustum of a cone that has been cut obliquely (Fig. 14/19)

Divide the plan into twelve equal sectors and number them from 1 to 12. Project these down to the F.E. and draw lines from each number to the apex A. You can see where each of these lines crosses the oblique top of the frustum. Now draw the basic development of the cone and number each sector from 1 to 12 and draw a line between each number and the apex A.

The lines A1 and A7 on the F.E. are the true length of the slant height of the cone. In fact, all of the lines from A to each number are equal in length but, on the F.E., lines A2 to A6 and A8 to A12 are shorter than A1 and A7 because they are sloping 'inwards' towards A. The true lengths from A to the oblique top of the frustum on these lines is found by projecting horizontally across to the line A1. Here, the true length can be swung round with compasses to its respective sector and the resulting series of points joined together with a neat curve.

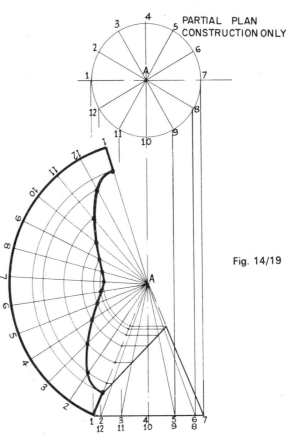

3RD ANGLE PROJECTION

PARTIAL PLAN
CONSTRUCTION ONLY

Fig. 14/19

1ST ANGLE PROJECTION

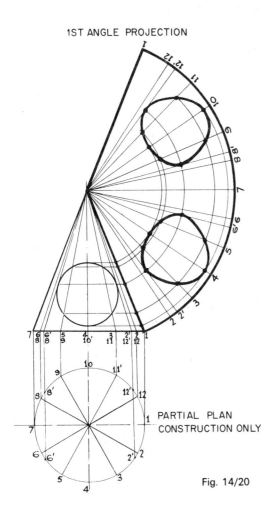

PARTIAL PLAN
CONSTRUCTION ONLY

Fig. 14/20

To develop a cone that has a cylindrical hole cut right through (Fig. 14/20)

This development, with one addition, is similar to the last example. Divide the plan into twelve sectors, number them and project them up to the F.E. Draw the basic development and mark and number the sectors on this development. The points where the lines joining the apex to numbers 3, 4, 5, 9, 10 and 11 cross the hole are projected horizontally to the side of the cone. They are then swung round to meet their respective sectors on the development.

There are four more points that need to be plotted. These are found by drawing tangents to the hole from the apex to meet the base in 6'8' and 2'12'. Project these points down to the plan so that their distances from the nearest sector line can be measured with dividers and transferred to the development. The point of tangency is then projected onto the development from the F.E. in the usual way.

155

To develop an oblique cone (Fig. 14/21) (An oblique cone, unlike a right cone, does not have its apex directly above the centre of its base.)

In the plan, Fig. 14/21, the base is divided into twelve equal sectors. These sectors are numbered and lines are drawn from each sector to the apex A. The true length of each of these lines is found by swinging them round in the plan to 2', 3', etc. and projecting up to the F.E. to give the true lengths A1 A2', A3', etc.

With centre A, arcs are drawn with radii equal to these true lengths.

The distance R on the plan (approximately 1/12 circumference of the base circle) is stepped from arc to arc, starting from point 7. The points are then joined together with a neat curve.

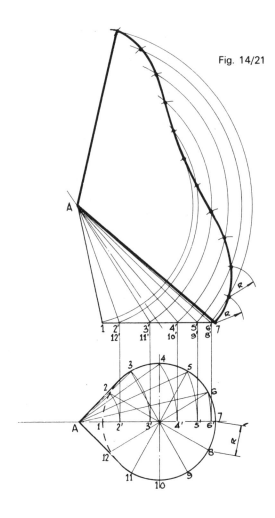

Fig. 14/21

Exercises 14

(All questions originally set in Imperial units)

1. Two views and an isometric view of a cement mixer cover are given in Fig. 1. Using a scale of 1/12, draw the two given views and add an end elevation. Then, using the same scale, draw the development of the sheet steel needed to make this cover.
Southern Regional Examination Board

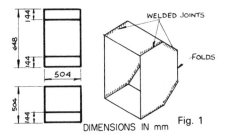

WELDED JOINTS

FOLDS

504

DIMENSIONS IN mm

Fig. 1

2. A small scoop is to be made to the dimensions given in the elevation, Fig. 2. Draw the development of the shape of the metal required for the body of the scoop with the joint on AB. Ignore the thickness of the metal and do not allow for any overlap.
Middlesex Regional Examining Board

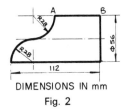

DIMENSIONS IN mm

Fig. 2

3. The Plan is given in Fig. 3 for the bucket seat for a 'Go-Kart', also the shape of the piece of plywood which is bent to form the back of the seat. Ignoring thicknesses draw:
(a) a Plan; (b) the development shown; (c) a Side Elevation of the assembled seat looking from A.
South-East Regional Examinations

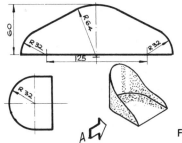

DIMENSIONS IN mm

Fig. 3

4. Fig. 4 shows the plan and elevation of a tin-plate dish. Draw the given views and construct a development of the dish showing each side joined to a square base. The plan of the base should be part of the development.
Middlesex Regional Examining Board

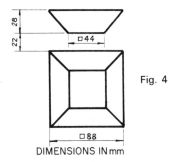

Fig. 4

DIMENSIONS IN mm

5. Fig. 5 shows a smoke stack and apron, both of which are circular. Draw, to a scale of 1/12, a development of the *apron only*, ignoring the thickness of the metal.
East Anglian Examinations Board

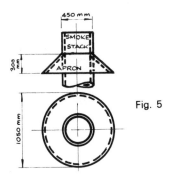

Fig. 5

6. Fig. 6 shows three pipes, each of 50 mm diameter and of negligible thickness, with their axes in the same plane and forming a bend through 90°. Draw:
(a) the given view, and (b) the development of pipe K, using TT as the joint line.
Associated Examining Board

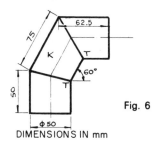

Fig. 6

DIMENSIONS IN mm

7. A front elevation of the body of a small metal jug is given in Fig. 7. Draw, *full size,* the following:
(a) the front view of the body as shown; (b) the side view of the body looking in the direction of arrow S; (c) the development of the body with the joint along AB.
South-East Regional Examinations Board (See Ch. 11 for information not in Ch. 14)

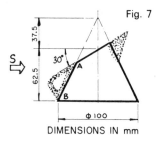

Fig. 7

DIMENSIONS IN mm

8. Two views of a solid are given in Fig. 8. Determine the development of the curved surface of the solid.
Oxford and Cambridge Schools Examination Board

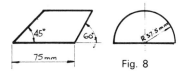

Fig. 8

9. The plan and elevation of a thin metal sheet are shown in first-angle projection in Fig. 9. A bar D of diameter 38 mm is placed on the plate which is then tightly wrapped round the bar so that edges A and B of the plate meet along a line at X. Draw a plan view of the wrapped plate when looking in the direction of arrow N, assuming that the bar has been removed.
Cambridge Local Examinations

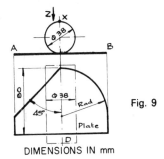

Fig. 9

DIMENSIONS IN mm

10. Fig. 10 shows the elevation and partly finished plan of a truncated regular pentagonal pyramid in first angle projection.
(a) Complete the plan view. (b) Develop the surface area of the sloping sides.
Cambridge Local Examinations

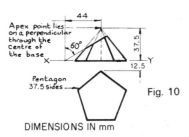

Apex point lies on a perpendicular through the centre of the base

60°

44

37.5

X — Y

12.5

Pentagon 37.5 sides

Fig. 10

DIMENSIONS IN mm

11. Fig. 11 shows two views of an oblique regular hexagonal pyramid. Draw, full size:
(a) the given views, and (b) the development of the sloping faces only, taking 'AG' as the joint line. Show the development in one piece.
Associated Examining Board

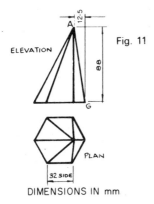

ELEVATION

A

12.5

88

G

Fig. 11

PLAN

32 SIDE

DIMENSIONS IN mm

12. Draw the development of the curved side of the frustum of the cone, shown in Fig. 12, below the cutting plane RST. Take JJ as the joint line for the development.
Associated Examining Board

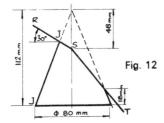

R

30°

J

S

48 mm

112 mm

J

18 mm

Ø 80 mm

T

Fig. 12

13. Make an accurate development of the sheet metal adaptor piece which is part of the surface of a right circular cone as shown in Fig. 13. The seam is at the position marked GH.
Cambridge Local Examinations

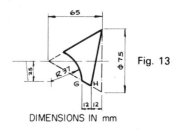

65

Ø 75

25

R 37

G H

12 12

Fig. 13

DIMENSIONS IN mm

158

15

Further problems in loci

We now come to an interesting set of curves: the cycloidal curves, the involute, the Archimedean spiral and the helix.

THE CYCLOID

The cycloid is the locus of a point on the circumference of a circle as the circle rolls, without slipping, along a straight line.

The approach to plotting a cycloid, as with all problems with loci, is to break down the total movement into a convenient number of parts and consider the conditions at each particular part Fig. 15/1. We have found, when considering the circle, that twelve is the most convenient number of divisions. The total distance that the circle will travel in 1 revolution is ΠD, the circumference, and this distance is also divided into twelve equal parts. When the circle rolls along the line. the locus of the centre will be a line parallel to the base line and the exact position of the centre will, in turn, be directly above each of the divisions marked off.

If a point P, on the circumference, is now considered, then after the circle has rotated 1/12 of a revolution point P is somewhere along the line $P_1 P_{11}$. The distance from P to the centre of the circle is still the radius and thus, if the inter-section of the line $P_1 P_{11}$ and the radius of the circle, marked off from the new position of the centre O_1, is plotted, then this must be the position of the point P after 1/12 of a revolution.

After 1/6 rev. the position of P is the intersection of the line $P_2 P_{10}$ and the radius, marked off from O_2. This is repeated for the twelve divisions.

Fig. 15/1 also shows the beginning of a second cycloid and it can be seen that the change from one cycloid to another is sudden. If any locus is plotted and has an instantaneous change of shape it indicates that there is a cessation of movement. Anything that has mass cannot change direction suddenly without first ceasing to move. The point of the circle actually in contact with the line is stationary.

This raises the interesting point that, theoretically, a motor car tyre is not moving at all when it is in contact with the road. This is not true in practice, since the contact between the road and tyre is not a point contact, but it does explain why tyres last much longer than would be expected.

At the top of the cycloid, between points 5 and 7, the point P is travelling nearly twice the distance that the centre moves in 1/12 rev. Thus, a jet car travelling at 800 km/h has points on the rim of the tyre moving up to 1,600 km/h—faster than the speed of sound.

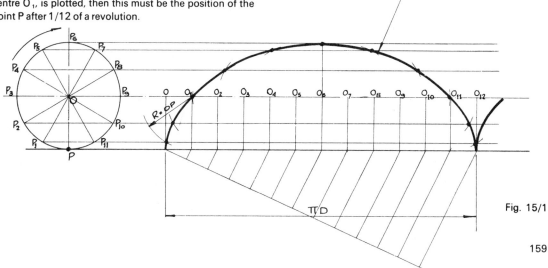

CYCLOID

Fig. 15/1

The tangent and normal to the cycloid (Fig. 15/2)
From the point P, where you wish to draw the normal and
the tangent, draw an arc whose radius is the same as the
rolling circle, to cut the centre line in O.

With centre O, draw the rolling circle to touch the base
line in Q.

PQ is the normal. The tangent is found by erecting a
line at 90° to the normal.

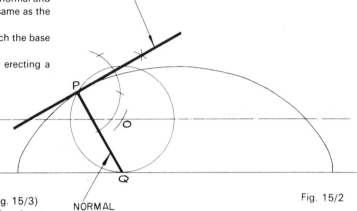

Fig. 15/2

The epi-cycloid and the hypo-cycloid (Fig. 15/3)
The epi-cycloid is the locus of a point on the circumference of a circle when it rolls, without slipping, along the
outside of a circular arc.

A hypo-cycloid is the locus of a point on the circumference of a circle when the circle rolls, without slipping,
along the *inside* of a circular arc.

The constructions for plotting these curves are very
similar to those used for plotting the cycloid.

The circumference of the rolling circle must be plotted
along the arc of the base circle. It is possible to calculate
this circumference and to plot it along the arc, but this is
fairly complicated and it is sufficiently accurate to measure
1/12 of the circumference of the rolling circle and step
this out 12 times, with dividers, along the base arc.

The remaining construction is similar to that used for
the cycloid. The technique is still to plot the intersection
of the line drawn parallel to the base, in this case another
arc with centre C, and the radius of the rolling circle from
its position after 1/12, 1/6, 1/4 revolutions, etc.

The main point to watch is that the locus of the centre
is no longer coincident with the line $P_3 P_9$ as it was for the
cycloid.

The epi-cycloid and the hypo-cycloid form the basis
for the shape of some gear teeth, although cycloidal
gear teeth have now generally been superseded by gear
teeth based on the involute.

**The tangent and normal to the epi-cycloid and
hypo-cycloid**
The method of obtaining the tangent and normal of an
epi-cycloid or a hypo-cycloid is exactly the same as for a
cycloid.

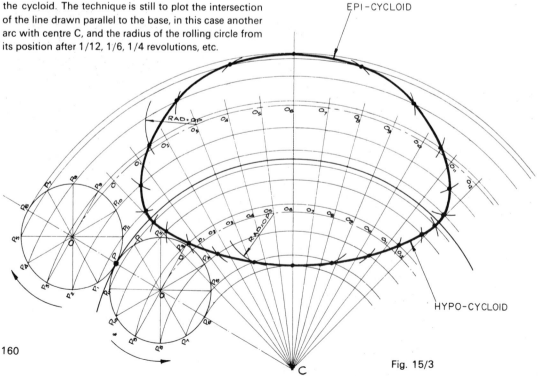

Fig. 15/3

THE TROCHOID

A trochoid is the locus of a point, not on the circumference of a circle but attached to it, when the circle rolls, without slipping, along a straight line.

Again, the technique is similar to that used for plotting the cycloid. The main difference in this case is that the positions of the line $P_1 P_{11}$, $P_2 P_{10}$, etc. are dependent upon the distance of P to the centre O of the rolling circle—not on the radius of the rolling circle as before.

This distance PO is also the radius to set on your compasses when plotting the intersections of that radius and the lines $P_1 P_{11}$, $P_2 P_{10}$, etc.

If P is *outside* the circumference of the rolling circle the curve produced is called a superior trochoid, Fig. 15/4.

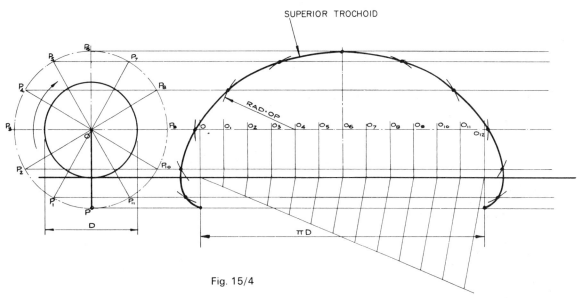

Fig. 15/4

If P is *inside* the circumference of the rolling circle the curve produced is called an inferior trochoid, Fig. 15/5.

The trochoid has relevance to naval architects. Certain inverted trochoids approximate to the profile of waves and therefore have applications in hull design.

The superior trochoid is the locus of the point on the outside rim of a locomotive wheel. It can be seen from Fig. 15/4 that at the beginning of a revolution this point is actually moving backwards. Thus, however quickly a locomotive is moving, some part of the wheel is moving back towards where it came from.

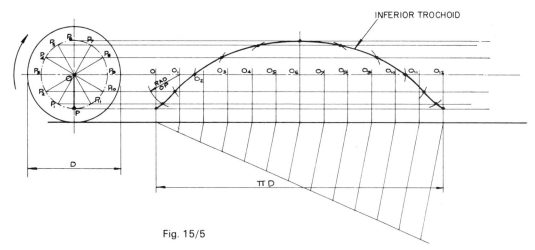

Fig. 15/5

161

THE INVOLUTE

There are several definitions for the involute, none being particularly easy to follow.

An involute is the locus of a point, initially on a base circle, which moves so that its straight line distance, along a tangent to the circle, to the tangential point of contact, is equal to the distance along the arc of the circle from the initial point to the instant point of tangency.

Alternatively, the involute is the locus of a point on a straight line when the straight line rolls round the circumference of a circle without slipping.

The involute is best visualized as the path traced out by the end of a piece of cotton when the cotton is unrolled from its reel.

A quick, but slightly inaccurate, method of plotting an involute is to divide the base circle into 12 parts and draw tangents from the twelve circumferential divisions, Fig. 15/6. Measure 1/12 of the circumference with dividers. When the line has unrolled 1/12 of the circumference, this distance is stepped out from the tangential point. When the line has unrolled 1/6 of the circumference, the dividers are stepped out twice. When 1/4 has unrolled the dividers are stepped out three times, etc. When all twelve points have been plotted they are joined together with a neat freehand curve.

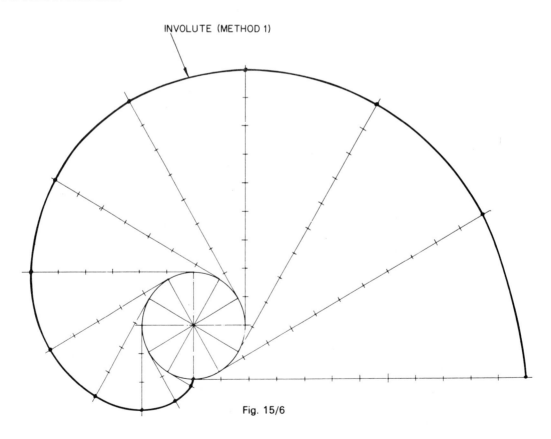

INVOLUTE (METHOD 1)

Fig. 15/6

A more accurate method is to calculate the circumference and lay out this length from a point on the base circle, Fig. 15/7. Divide the length into 12 equal parts and use compasses to swing the respective divisions to their intersections with the tangents.

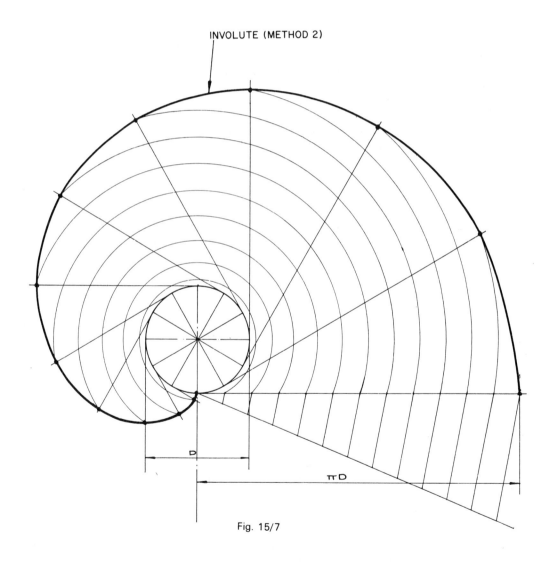

INVOLUTE (METHOD 2)

D

π D

Fig. 15/7

The normal and tangent to an involute

The construction for the normal, and hence the tangent, to an involute relies on the construction of a tangent from a point to a circle and for this reason occasionally appears in G.C.E. 'O' level papers.

The construction, shown in Chapter 5 and Fig. 15/8, is to draw a line from the point on the involute to the centre of the base circle and bisect it. This gives the centre of a semi-circle, radius half the length of the line, which crosses the base circle at point T. The normal is then drawn from the point on the involute through the point T. The tangent is found by erecting a perpendicular to the normal from T.

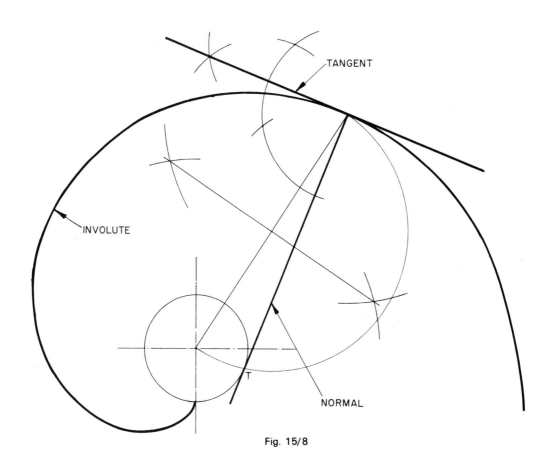

TANGENT

INVOLUTE

T

NORMAL

Fig. 15/8

THE ARCHIMEDEAN SPIRAL

The Archimedean spiral is the locus of a point which moves away from another fixed point at uniform linear velocity and uniform angular velocity.

It may also be considered to be the locus of a point moving at constant speed along a line when the line rotates about a fixed point at constant speed.

Since both the linear and angular speeds are constant, the only rule for plotting an Archimedean spiral is that the linear and angular distances moved through must both be divided into the same number of equal parts. The most convenient number of equal parts is 12 and if one convolution (when dealing with spirals a movement through 360° is called a convolution as distinct from a revolution) is to be plotted, then the linear distance moved through is divided into 12 equal parts and the 360° into 30° intervals, Fig. 15/9. The linear divisions may then be swung round to intersect with the respective angular divisions.

If more than one convolution is to be drawn, then, although the number of angular divisions remains at twelve, the linear divisions must be divided into the appropriate multiple. Thus, if two convolutions are to be drawn, there will be 24 linear divisions, etc.

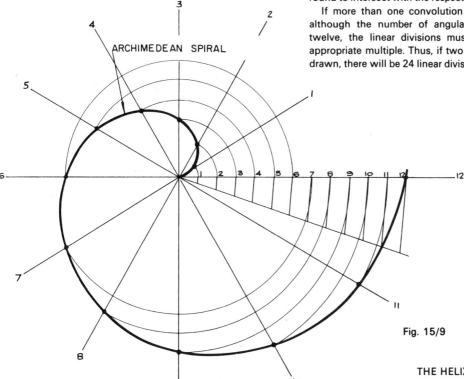

ARCHIMEDEAN SPIRAL

Fig. 15/9

THE HELIX

The helix is the locus of a point which moves round a cylinder at constant velocity whilst advancing along the cylinder at constant velocity.

It is the curve obtained when a piece of string is wound round a cylinder.

It is also the curve generated when turning on a lathe. Normally, the rate of advance along the piece of work is so small that it is indistinguishable as a helix, but it is quite easily seen when cutting a screw thread.

The distance that the point moves along the cylinder in one complete revolution is called the *pitch*.

The construction of the helix is simple, Fig. 15/10. The movement round and along the cylinder is constant and so, for a fixed period, say one complete revolution, the two movements are divided into the same number of equal parts. Thus 1/12 rev. will coincide with 1/12 pitch, 1/6 rev. will coincide with 1/6 pitch, etc.

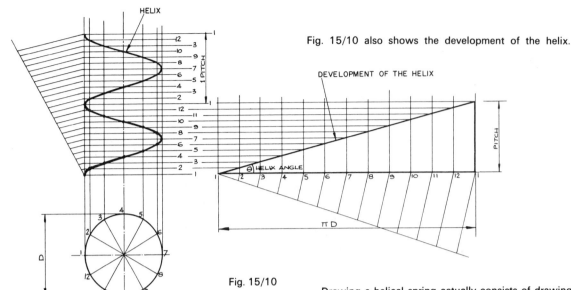

Fig. 15/10 also shows the development of the helix.

Fig. 15/10

Coiled springs

Most coiled springs are formed on a cylinder and are, therefore, helical. They are, in fact, more often called helical springs than coiled springs. If the spring is to be used in tension, the coils will be close together to allow the spring to stretch. This is the spring that you will see on spring balances in the science lab. If the spring is to be in compression, the coils will be further apart. These springs can be seen on the suspension of many modern cars, particularly on the front suspension.

Drawing a helical spring actually consists of drawing two helices, one within another. Although the diameters of the helices differ, their pitch must be the same. Once the points are plotted it is just a question of sorting out which parts of the helices can be seen and which parts are hidden by the thickness of the wire.

For clarity, the thickness of the wire in Fig. 15/11 is 1/4 the pitch of the helix, but if it wasn't a convenient fraction, it would be necessary to set out the pitch twice. The distance between the two pitches would be the thickness of the wire.

COILED SPRINGS

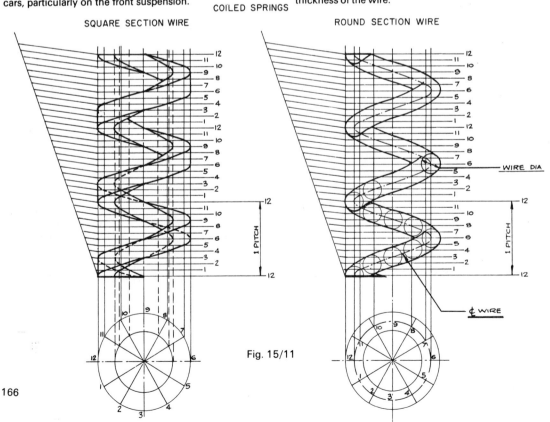

Fig. 15/11

Screw thread projection

A screw thread is helical. Unless the screw thread is drawn at a large scale, it is rarely drawn as a helix— except as an exercise in drawing helices!

A good example is to draw a screw thread with a square section. This is exactly the same construction as the coiled spring except that the central core hides much of the construction.

A right-hand screw thread is illustrated in Fig. 15/12. To draw a left-hand screw thread merely plot the ascending points from right to left instead of from left to right.

Sometimes a double, triple or even a quadruple start thread is seen, particularly on the caps of some containers

A SINGLE START SQUARE THREAD

Fig. 15/12

A TRIPLE START SQUARE THREAD

Fig. 15/13

where the top needs to be taken off quickly. A multiple start thread is also seen on the starter pinion of motor cars. Multiple start screw threads are used where rapid advancement along a shaft is required. When plotting a double start screw thread, two helices are plotted on the same pitch. The first helix starts at point 1 and the second at point 7. If a triple start screw thread is plotted, the starts are points 1, 9 and 5, Fig. 15/13. If a quadruple start thread is plotted, the starts are points 1, 10, 7 and 4.

Exercises 15

(All questions originally set in Imperial units)

1. Fig. 1. shows a circular wheel 50 mm in diameter with a point P attached to its periphery. The wheel rolls without slipping along a perfectly straight track whilst remaining in the same plane.
   Plot the path of point P for one-half revolution of the wheel on the track. Construct also the normal and tangent to the curve at the position reached after one-third of a revolution of the wheel.
   Cambridge Local Examinations

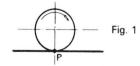

Fig. 1

2. The views in Fig. 2 represent two discs which roll along AB. Both discs start at the same point and roll in the same direction. Plot the curves for the movement of points *p* and *q* and state the perpendicular height of *p* above AB where *q* again coincides with the line AB.
Southern Universities' Joint Board

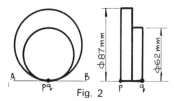

Fig. 2

3. A wheel of 62 mm diameter rolls without slipping along a straight path. Plot the locus of a point P on the rim of the wheel and initially in contact with the path, for one half revolution of the wheel along the path. Also construct the tangent, normal and centre of curvature at the position reached by the point P after one quarter revolution of the wheel along the path.
Cambridge Local Examinations

4. The driving wheels and coupling rod of a locomotive are shown to a reduced scale in Fig. 3. Draw the locus of any point P on the link AB for one revolution of the driving wheels along the track.
University of London School Examinations

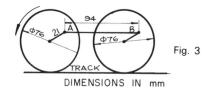

Fig. 3

DIMENSIONS IN mm

5. A piece of string AB, shown in Fig. 4, is wrapped around the cylinder, centre O, in a clockwise direction. The length of the string is equal to the circumference of the cylinder.
   (a) Show, by calculation, the length of the string, correct to the nearest 1 mm, taking $\Pi = 3.14$.
   (b) Plot the path of the end B of the string as it is wrapped round the cylinder, keeping the string taut.
   (c) Name the curve you have drawn.
Middlesex Regional Examining Board

Fig. 4

6. A cylinder is 48 mm diameter and a piece of string is equal in length to the circumference. One end of the string is attached to a point on the cylinder.
   (a) Draw the path of the free end of the string when it is wound round the cylinder in a plane perpendicular to the axis of the cylinder.

(b) In block letters, name the curve produced.
   (c) From a point 56 mm chord length from the end of the curve (i.e. the free end of the string) construct a tangent to the circle representing the cylinder.
Southern Universities' Joint Board

7. A circle 50 mm diameter rests on a horizontal line. Construct the involute to this circle, making the last point on the curve 2 Π mm from the point at which the circle makes contact with the horizontal line.
Cambridge Local Examinations

8. P, O and Q are three points in that order on a straight line so that PO = 34 mm and OQ = 21 mm. O is the pole of an Archimedean spiral. Q is the nearest point on the curve and P another point on the first convolution of the curve. Draw the Archimedean spiral showing two convolutions.
Southern Universities' Joint Board

9. Draw two convolutions of an Archimedean spiral such that in two revolutions the radius increases from 18 mm to 76 mm.
Oxford Local Examinations

10. A piece of cotton is wrapped around the cylinder shown in Fig. 5. The cotton starts at C and after one turn passes through D, forming a helix. The start of the helix is shown in the figure. Construct the helix, showing hidden detail.
Middlesex Regional Examining Board

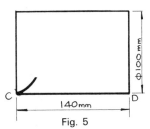

Fig. 5

11. A cylinder, made of transparent material, 88 mm O/D, 50 mm I/D, and 126 mm long, has its axis parallel to the V.P. Two helical lines marked on its curved surface—one on the outside and the other on the inside—have a common pitch of 63 mm.
    Draw the elevation of the cylinder, showing both helices starting from the same radial line and completing two turns.
Associated Examining Board

12. Draw a longitudinal elevation, accurately projected, showing two turns of a helical spring. The spring is of 100 mm outside diameter, the pitch of the coils is 62 mm and the spring material is of 10 mm diameter.
Cambridge Local Examinations

16
Freehand sketching

The ability to sketch neatly and accurately is one of the most useful attributes that a draughtsman can have. Freehand drawing is done on many occasions: to explain a piece of design quickly to a colleague; to develop a design (see Fig. 16/10); and even to draw a map showing someone how to get from one place to another.

Technical sketching is a disciplined form of art. Objects must be drawn exactly as they are seen, not as one would like to see them. Neat, accurate sketches are only achieved after plenty of practice, but there are some guiding rules.

Most engineering components have outlines composed of straight lines, circles and circular arcs: if you can sketch these accurately, you are halfway towards producing good sketches. You may find the method illustrated in Fig. 16/1 a help. When drawing straight lines, as on the left, rest the weight of your hand on the backs of your fingers. When drawing curved lines, as on the right, rest the weight on that part of your hand between the knuckle of your little finger and your wrist. This provides a pivot about which to swing your pencil. Always keep your hand on the *inside* of the curve, even if it means moving the paper around.

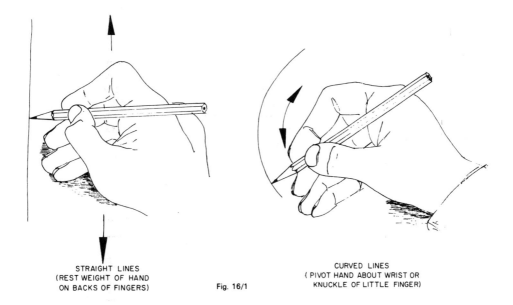

STRAIGHT LINES
(REST WEIGHT OF HAND
ON BACKS OF FINGERS)

Fig. 16/1

CURVED LINES
(PIVOT HAND ABOUT WRIST OR
KNUCKLE OF LITTLE FINGER)

Pictorial Sketching

Freehand pictorial sketching looks very much like isometric drawing. Circles appear as ellipses and lines are drawn at approximately 30°. Circles have been sketched onto an isometric cube in Fig. 16/2. You can see how these same ellipses appear on sketches of a round bar material.

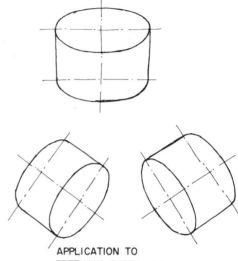

CIRCLES SKETCHED ON
AN ISOMETRIC CUBE

APPLICATION TO
FREEHAND SKETCHING
OF CYLINDRICAL OBJECTS

Fig. 16/2

When sketching, you may find it an advantage to draw a faint 'box' first and draw in the ellipses afterwards. With practice you should find that you can draw quite a good ellipse if you mark out its centre lines and the major and minor axes.

Although drawing is a continuous process, the work can be divided into three basic stages.

Stage 1 Construction

This should be done with a hard pencil (6H), used lightly, and the strokes with the pencil should be rapid. Slow movements produce wavy, uncertain lines. Since these constructed lines are very faint, errors can easily be erased.

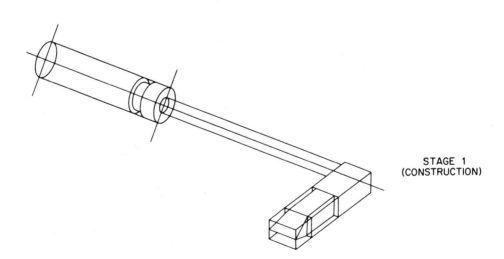

STAGE 1
(CONSTRUCTION)

Stage 2 Lining in

Carefully line in with a soft pencil (HB), following the construction lines drawn in stage 1.

The completion of stage 2 should give a drawing which shows all the details and you may decide, particularly in an examination, not to proceed to stage 3.

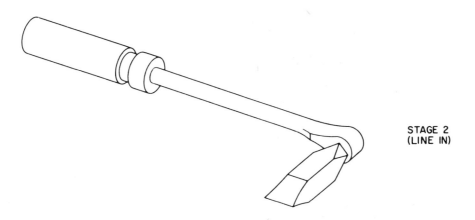

STAGE 2
(LINE IN)

Stage 3 Shading

Shading brings a drawing to life. It is not necessary on most sketches, and in some cases it may tend to hide details which need to be seen. If the drawings are to be displayed, however, some shading should certainly be done.

Shading is done with a soft pencil (HB). It is very easy to overshade, so be careful. For the smooth merging of shading, the dry tip of a finger can be gently rubbed over the area.

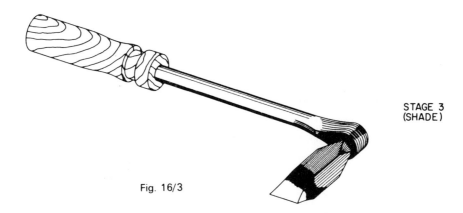

STAGE 3
(SHADE)

Fig. 16/3

Figs. 16/3,4,5,6 and 7 are examples of freehand pictorial sketching.

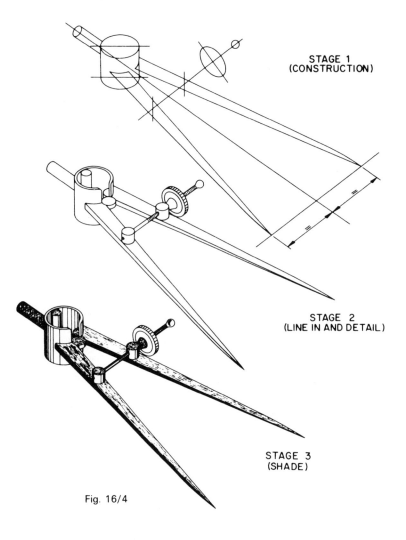

STAGE 1
(CONSTRUCTION)

STAGE 2
(LINE IN AND DETAIL)

STAGE 3
(SHADE)

Fig. 16/4

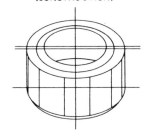

STAGE 1
(CONSTRUCTION)

STAGE 2
(LINE IN AND DETAIL)

STAGE 3
(SHADE)

Fig. 16/5

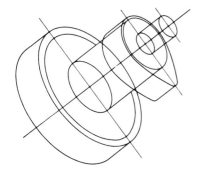

STAGE 1
(CONSTRUCTION)

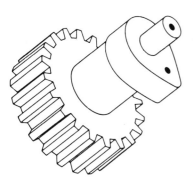

STAGE 2
(LINE IN AND DETAIL)

Fig. 16/6

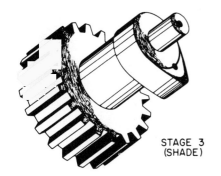

STAGE 3
(SHADE)

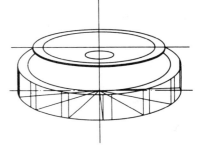

STAGE 1
(CONSTRUCTION)

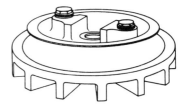

STAGE 2
(LINE IN AND DETAIL)

STAGE 3
(SHADE)

Fig. 16/7

Sketching in Orthographic Projection

More detail can be seen on an orthographic drawing than on an isometric, mainly because more than one view is drawn. For this reason it is often advantageous for a draughtsman to make an orthographic sketch.

The views should be drawn in the conventional orthographic positions, i.e., in 3rd angle projection, a F.E., a plan above the F.E. and an E.E. to the left or right of the F.E.; in 1st angle projection, a F.E., a plan below the F.E. and an E.E. to the left or right of the F.E. These views should be linked together with projection lines.

1ˢᵀ ANGLE PROJECTION

SECTION X–X

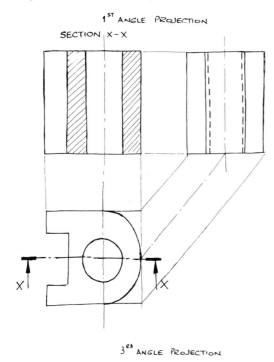

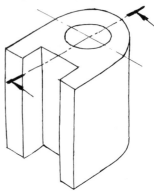

Fig. 16/8

Figs. 16/8 and 16/9 show pictorial and orthographic sketches of two engineering components.

3ᴿᴰ ANGLE PROJECTION

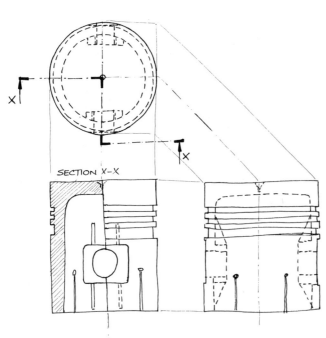

SECTION X–X

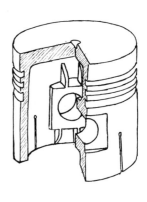

Fig. 16/9

Circles are difficult to draw freehand but you can use your hand as a compass. Hold your pencil upright and, using your little finger as a compass 'point', revolve the paper keeping your hand quite still.

Fig. 16/10 shows how a draughtsman might use sketching to aid a piece of design. He wishes to design a small hand vice.

First he makes a freehand pictorial sketch so that he can see what the vice will look like. He also makes a few notes about some details of the vice.

He then makes an orthographic sketch. This shows much more detail and he makes some more notes.

This is quite a simple vice and he may now feel that he is ready to make the detailed drawings. If it was more complicated he might make a few more sketches showing even more detail. Details of one of the legs of the vice are shown.

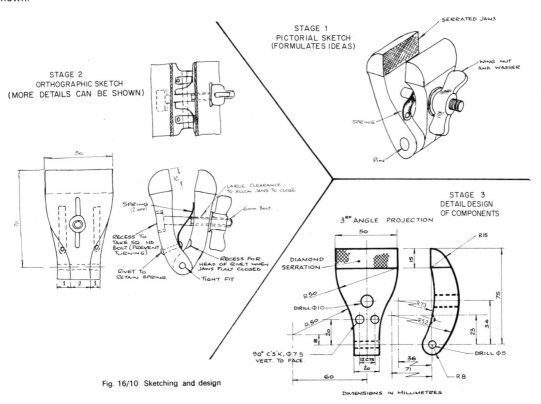

Fig. 16/10 Sketching and design

175

17

Some more problems solved by drawing

This chapter introduces the student to some more drawing techniques. It should be emphasized that the topics are only introduced; all of them can be studied in much greater depth and any solutions offered in this chapter will apply to simple problems only.

AREAS OF IRREGULAR SHAPES

It is possible to find, by drawing, the area of an irregular shape. The technique does not give an exact answer but, carefully used, can provide a reasonable answer. Look at Fig. 17/1. The shape is trapezoidal with irregular end lines. A centre line has been drawn and you can see that

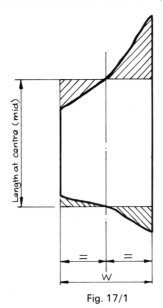

Fig. 17/1

the shaded triangles at the top and bottom are, in their pairs, approximately the same in area. Thus, the approximate area of the whole figure is the width W multiplied by the height at the centre. That height is called the mid-ordinate and the whole technique is called the *mid-ordinate rule*.

Fig. 17/2 shows how it is applied to a larger figure.

The figure is divided into a number of equal strips (width W), in this case 8. The more strips that are drawn (within reason) the greater the accuracy of the final calculation. The centre line of each strip (the mid-ordinate) is drawn. Each of these lengths is measured and the area of the figure is given by

$$W(O_1 + O_2 + O_3 + O_4 + O_5 + O_6 + O_7 + O_8)$$

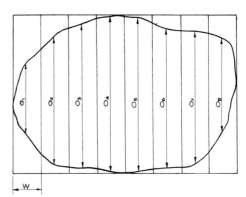

Fig. 17/2

A sample of this technique is shown in Fig. 17/3. The curve shown is a sin curve, the curve that emerges if you plot the values of the sin of all the angles from 0° to 90°.

The area is the product of the width of each strip and the sum of all the mid-ordinates.

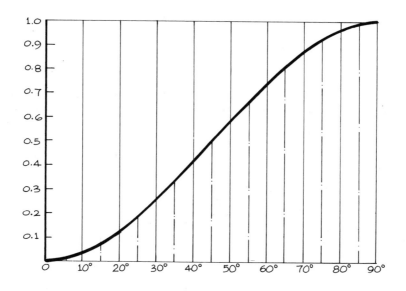

Fig. 17/3

RESOLUTION OF FORCES

All machinery, however simple. has forces acting on its parts. Buildings have forces acting on them; forces produced by the weight of the building itself, the weight of the things inside it and by the wind pushing against it. An understanding of how these forces act and how they affect design is essential to a good draughtsman.

You must first understand the difference between stable and unstable forces and then study the effects of the unstable ones. The two men indulging in indian wrestling in Fig. 17/4 are applying force. As long as the forces are equal they will remain in the position shown. When one begins to apply more force than the other the forces become unstable and the other has his hand forced back onto the table.

STABLE FORCES

Some forces acting on a point are shown in Fig. 17/5. The forces on the left are in line, they are equal and opposite and so the forces are stable. The centre forces are in line and opposite but one is larger than the other. Therefore the larger force will push the smaller force back. The forces on the right are equal in size but are not in line. The effect will be for the point to move in the direction shown.

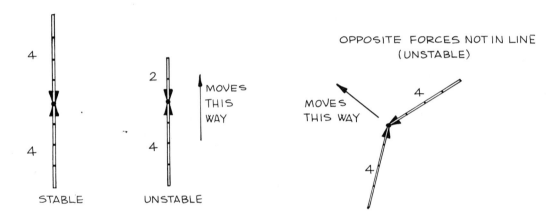

Fig. 17/5   Forces acting on a point

Some forces acting on a beam are shown in Fig. 17/6. The forces on the left are in line and opposite and so the forces are stable. The four forces on the second beam are also equal and opposite and so, although they are acting in pairs, the whole set of forces are stable. The forces acting on the third example are equal but not opposite.

The effect of these forces acting in the way they are placed will be to rotate the beam as shown. This kind of force is important and is called a *couple*. The final set of forces on the right are also stable. The two forces under the beam add up to the same total as the force above the beam. They are spaced at equal distance from the force above the beam and thus the whole system is stable.

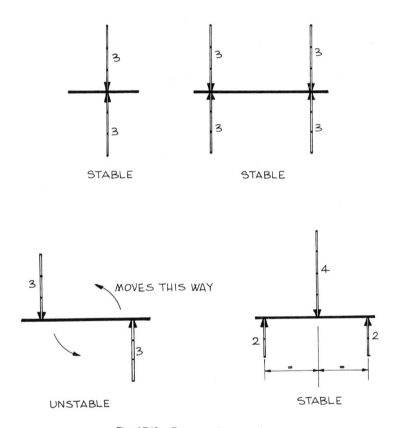

Fig. 17/6   Forces acting on a beam

Calculating some of the forces acting on a beam

If a beam is loaded then, to stabilise it, forces have to be applied in opposition. These opposite forces are called *reactions*. Some simple examples are shown in Fig. 17/7. The loading is applied at a point and is therefore called a *point load*.

The left hand example has the reactions positioned at equal distances from the load and they will therefore be equal. In that case they must each be half the load.

The centre example has a load of 8 units* and one reaction is 5 units. For the loading and reaction system to be stable, the sum of the reactions must equal the loading. Therefore the reaction on the right (R_R) must be 3 units.

The left and right reactions are given in the example on the right. Since the sum of the reactions must equal the load, the load is 5 units.

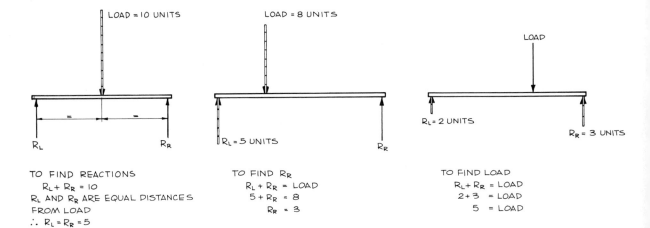

Fig. 17/7 Reactions in a point loaded beam

* Forces in the SI System are given in newtons (the force required to accelerate 1 kg at 1 metre per sec²). For simplicity the loads are called units in this chapter.

Fig. 17/8 shows two examples where the loads and reactions are known but the position of one reaction has to be calculated.

If the loaded system is stable then ALL the forces must balance. The reactions added together $(8+4)$ must equal the load (12). The couples must also be balanced out. The left hand couple, reaction times distance to the load (8×2) must equal the right hand couple $(4 \times x)$. Thus the distance can be worked out in the simple equation shown.

The example on the right is similar.

The force called a couple in this chapter is also called a *moment* or *bending moment* (because it tries to bend the beam) by engineers.

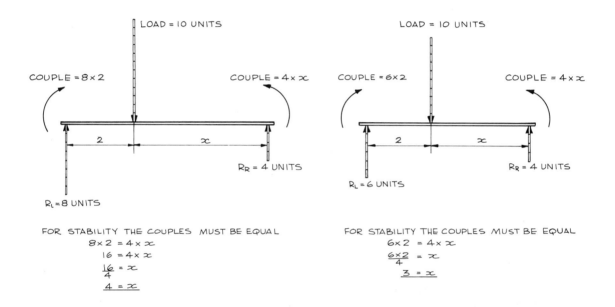

LOAD = 10 UNITS

COUPLE = 8 × 2

COUPLE = 4 × x

2

x

R_R = 4 UNITS

R_L = 8 UNITS

FOR STABILITY THE COUPLES MUST BE EQUAL

$$8 \times 2 = 4 \times x$$
$$16 = 4 \times x$$
$$\frac{16}{4} = x$$
$$4 = x$$

LOAD = 10 UNITS

COUPLE = 6 × 2

COUPLE = 4 × x

2

x

R_R = 4 UNITS

R_L = 6 UNITS

FOR STABILITY THE COUPLES MUST BE EQUAL

$$6 \times 2 = 4 \times x$$
$$\frac{6 \times 2}{4} = x$$
$$3 = x$$

Fig. 17/8 Finding the position of the reaction in a stable loaded beam

Two more examples are shown in Fig. 17/9. In this case the loading is not acting at a point but is evenly distributed along the beam. This could make calculations difficult; in fact we are able to change the loading. The total load is the load per metre multiplied by its length, ($6 \times 1 = 6$ units). The effect of this evenly distributed load on the beam is the same as if it were a point load acting at the centre. The lower left drawing in Fig. 17/9 shows how this looks. The reactions can now be easily calculated.

The example on the right also has the evenly distributed load changed into a point load acting at the centre of the beam. The size of the two reactions can be calculated with a simple simultaneous equation (shown underneath the diagram).

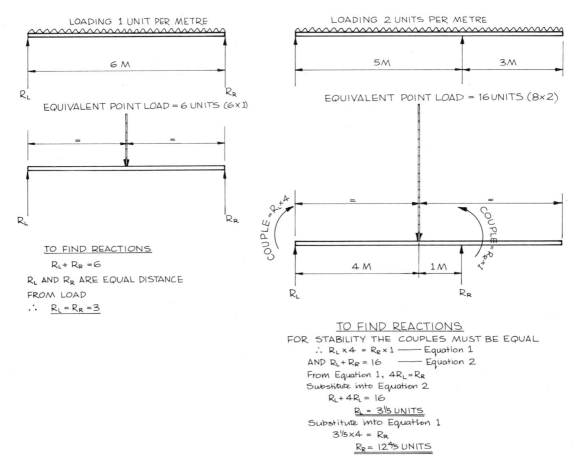

TO FIND REACTIONS

$R_L + R_R = 6$

R_L AND R_R ARE EQUAL DISTANCE FROM LOAD

$\therefore \quad R_L = R_R = 3$

TO FIND REACTIONS

FOR STABILITY THE COUPLES MUST BE EQUAL

$\therefore \quad R_L \times 4 = R_R \times 1$ —— Equation 1

AND $R_L + R_R = 16$ —— Equation 2

From Equation 1, $4R_L = R_R$

Substitute into Equation 2

$R_L + 4R_L = 16$

$R_L = 3\tfrac{1}{5}$ UNITS

Substitute into Equation 1

$3\tfrac{1}{5} \times 4 = R_R$

$R_R = 12\tfrac{4}{5}$ UNITS

Fig. 17/9 Reactions in an evenly distributed loaded beam

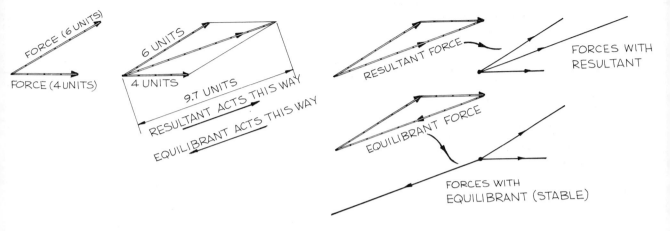

Forces acting at a point

When two or more forces act at or on a point it is useful to be able to change these forces and show how they could be replaced with a single force which acts in the same way. This force is called the *resultant force*. The force which would have to be applied to stabilize the system (by acting against and cancelling out the two or more forces) is called the *equilibrant force*. An example is shown in Fig. 17/10.

The two forces are of 6 units and 4 units. The resultant and equilibrant forces can be found by drawing lines parallel to these two forces and forming a parallelogram. The resultant and equilibrant forces are equal to the length of the longer diagonal of this parallelogram. An alternative is to draw triangles as shown in the figure. The result of these drawings is shown on the extreme right. The resultant force is the one that would have the same effects as the two forces if it replaced them; the equilibrant force acts against the two forces and makes the system stable.

Fig. 17/11 shows how to find the equilibrant force to three forces acting at a point. Draw a line parallel to force 1 and equal (in scale) to its force. From the end of this line draw a line parallel to force 2 and equal in scale) to its force. From the end of this line draw a line parallel to force 3 and equal (in scale) to its force. The line which closes the quadrilateral is equal to the size of the equilibrant force (to scale) and acts in the direction of the equilibrant force. It can be transferred back to the original drawing of the forces and stabilises the system.

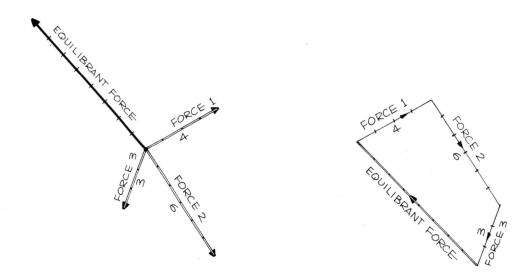

Fig. 17/11   Finding the equilibrant force to 3 forces meeting at a point

Fig. 17/12 shows how to find the equilibrant force to four forces acting at a point. Once again the forces are drawn to scale parallel to the way they are acting at the point. The forces are each taken in turn and are considered clockwise. The point to note on this example is that, although the lines cross each other the equilibrant force is still the one which closes the figure, the one that joins the end of the line representing the last force drawn to the beginning of the line representing the first force drawn.

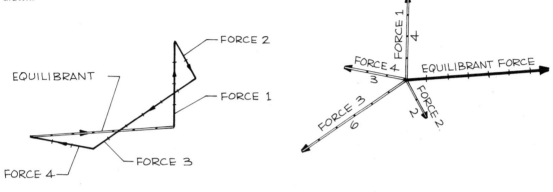

Fig. 17/12 Finding the equilibrant force to 4 forces meeting at a point

SIMPLE CAM DESIGN

Cams are used in machines to provide a controlled up and down movement. This movement is transmitted by means of a *follower*. A cam with a follower is shown in Fig. 17/13 in the maximum 'up' and 'down' positions. The difference between these two positions gives the *lift* of the cam. The shape of the cam, its *profile*, determines how the follower moves through its lift and *fall*.

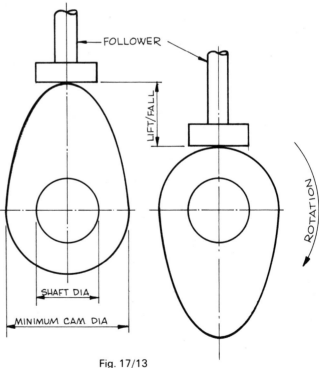

Fig. 17/13

185

A cam is designed to make a part of a machine move in a particular way. For example, cams are used to open and close the valves which control the petrol mixture going into and the exhaust gases coming out of an internal combustion engine. Obviously the valves must open at the right time and at the right speed and it is the cam that determines this. This control on the valve is exercised by the profile of the cam, determining how the follower lifts and falls. Three examples of types of lift and fall are shown in Fig. 17/14.

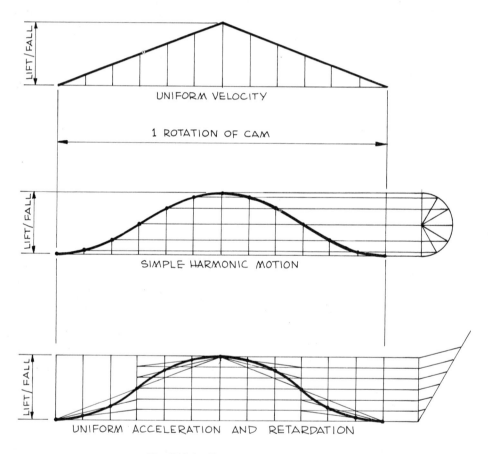

Fig. 17/14   Types of lift and fall

The top diagram shows uniform velocity. This is a graph of a point which is moving at the same velocity 'upward' to half way and then at the same velocity 'downward' to its starting point.

The centre diagram shows simple harmonic motion. This is the motion of a pendulum, starting with zero velocity, accelerating to a maximum and then decelerating to zero again (at the top of the curve) and then repeating the process back to the starting point. The curve is a sine curve and is plotted as shown.

The lower diagram shows uniform acceleration and retardation. In this case the acceleration is uniform to a point halfway up the lift and then retarding uniformly to the maximum lift. The process is then repeated in reverse back to the starting point. The curve is plotted as shown.

If a cam has to be designed it is drawn around a specification. This must state the dimensions, the lift/fall and the *performance*. The performance states how the follower is to behave throughout one rotation of the cam. The designer must first draw the performance curve to the given specification. An example is shown in Fig. 17/15. *Dwell* is a period when the follower is neither lifting or falling.

The performance curve is started by drawing a base line of 12 equal parts, the total representing one rotation of the cam. The lift/fall is then marked out, and the performance curve drawn. The base line of the performance curve is then projected across and, with the centre line of the cam, forms the top of the circle representing the minimum cam diameter. Once the centre of the cam has been found centre lines can be drawn at 30° intervals. The cam profile is plotted on these lines. Twelve points on the performance curve are then projected across to the centre line of the cam and then swung round with compasses to the intersecting points on the lines drawn at 30° intervals. If the cam is rotating clockwise the points 1 to 12 are marked out clockwise; if the rotation is anti-clockwise, as in this case, the points are marked out anti-clockwise.

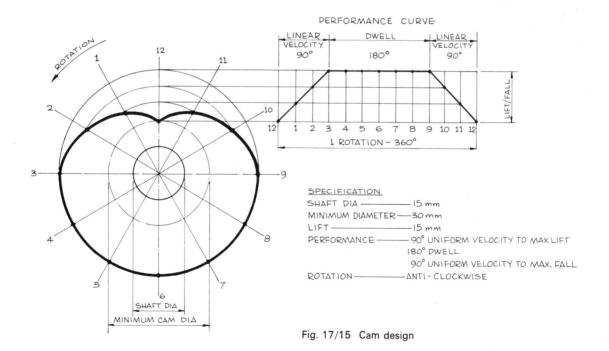

Fig. 17/15 Cam design

Fig. 17/16 shows another example of a cam design.
This is a more complicated profile than the previous
example but the method used to construct the profile
is the same. This cam rotates in a clockwise direction.

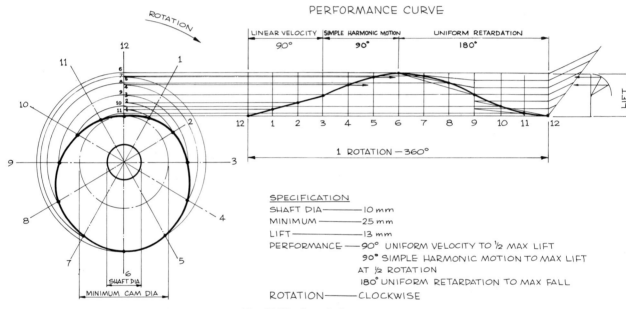

PERFORMANCE CURVE

SPECIFICATION
SHAFT DIA————10 mm
MINIMUM————25 mm
LIFT————————13 mm
PERFORMANCE——90° UNIFORM VELOCITY TO ½ MAX LIFT
             90° SIMPLE HARMONIC MOTION TO MAX LIFT
        AT ½ ROTATION
        180° UNIFORM RETARDATION TO MAX FALL
ROTATION————CLOCKWISE

Fig. 17/16 Cam design

188

Finally, three different types of followers are shown in Fig. 17/17. The left hand example is a knife follower. It can be used with a cam that has a part of its profile convex but it wears quickly. The centre example is a roller follower; this type of follower obviously reduces friction between the follower and the cam, something of a problem with the other types shown. The flat follower is an all purpose type which is widely used. It wears more slowly than a knife follower.

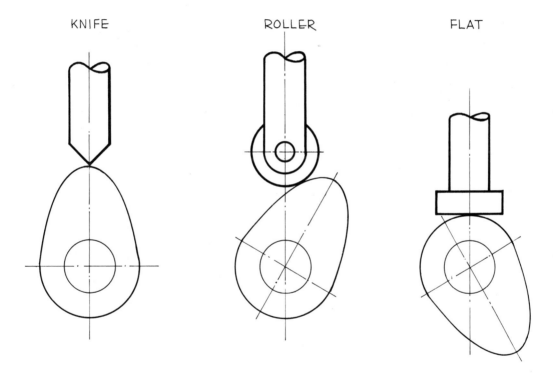

KNIFE ROLLER FLAT

Fig. 17/17   Three types of cam followers

Exercises 17

1. The velocity of a vehicle moving in a straight line from start to rest was recorded at intervals of 1 minute and these readings are shown in the table. Part of the velocity/time diagram is given in Fig. 1.

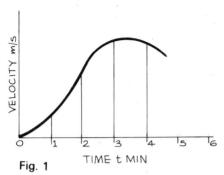

Fig. 1

Draw the complete diagram with a horizontal scale of 10 mm to 1 minute and vertical scale of 10 mm to 1 m/s. Using the mid ordinate rule, determine the average velocity of the vehicle and hence the total distance travelled in the 12 minutes.

Time t min	0	1	2	3	4	5	6
Vel. v m/s	0	1.4	5.4	9.2	8.6	7.9	9.0

Time t min		7	8	9	10	11	12
Vel. v m/s		10.2	11.5	12.6	9.4	5.2	0

Oxford Local Examinations

2. Fig. 2. shows part of an indicator diagram which was made during an engine test. Draw the complete diagram, *full size*, using the information given in the table and then by means of the mid-ordinate rule determine the area of the diagram. Also, given that the area of the diagram can be given by the product

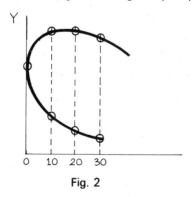

Fig. 2

of its length and average height, determine the average height and the average pressure if the ordinates represent the pressure to a scale of 1 mm to 60 kN/m².

OX(mm)	0	10	20	30	40	50
OYmax(mm)	54	79	82	78	66	54

OYmin(mm)	54	21	14	12	12	12
OX(mm)	60	70	80	90	100	

OYmax(mm)	45	39	33	29	21	
OYmin(mm)	12	12	12	13	21	

Oxford Local Examinations

3. Fig. 3 shows four simply loaded beams. Find the size of the reactions marked X.

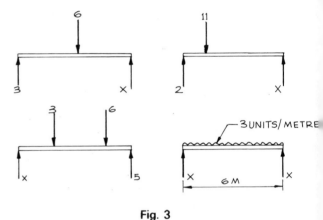

Fig. 3

4. Fig. 4 shows two simply loaded beams with their reactions. Find the dimensions marked x for the loading to be in equilibrium (the couples to be equal).

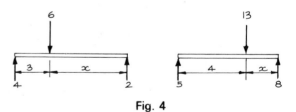

Fig. 4

5. Fig. 5 shows two beams loaded with evenly distributed loads. The positions of the reactions are shown. Find the size of the reactions.

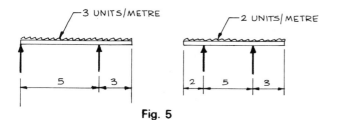

Fig. 5

6. Fig. 6 shows three examples of two forces acting at a point. Find the size of the resultant force for each example and measure the angular direction of the resultant force from the datum shown.

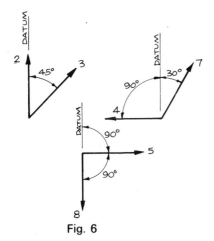

Fig. 6

7. Fig. 7 shows two examples of forces acting at a point. Find the size of the resultant force for both examples and measure and state the angular direction of the resultant force from the datum shown. From the same datum state the angular direction of the equilibrant force for both examples.

8. Plot the cam profile which meets the following specification:
Shaft diam.: 15 mm
Min. diam.: 25 mm
Lift: 12 mm
Performance: 90° uniform velocity to max. lift
90° dwell
180° uniform retardation to max. fall.
Rotation: Clockwise.
Your cam profile must be drawn *twice full size*.

9. Plot the cam profile which meets the following specifications:
Shaft diam.: 12.5 mm
Min. diam.: 30 mm
Lift: 12.5 mm
Performance: 60° dwell
90° simple harmonic motion to half lift
30° dwell
60° uniform acceleration to max. lift
120° uniform velocity to max. fall.
Rotation: anticlockwise
Your cam profile is to be drawn *twice full size*.

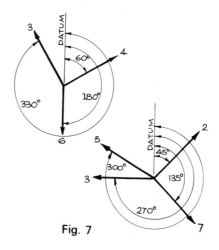

Fig. 7

Part II
Engineering drawing

18

Engineering drawing

Introduction

The first step a manufacturer must take when he wishes to make an article is to produce a drawing. First a designer will make a preliminary sketch and then a draughtsman will make a detailed drawing of the design. Since neither the designer nor the draughtsman will actually make the article, the drawings must be capable of being interpreted by the men in their workshops. These workshops may be sited a long way from the drawing office, even overseas, and so the drawings produced must be standardized so that anyone familiar with these standards could make the article required. Engineering drawing is, therefore, a language. In this modern age of rapid communication and international buying, from pins to complete atomic power stations, it is essential for the language to be international. This is the reason why you will often see symbols used on a drawing instead of words or abbreviations.

British Standard 308 gives the rules for engineering drawing and should be carefully studied by every prospective draughtsman.

The rest of this book tries to explain, within the framework of BS 308, the language of engineering drawing.

Type of projection

The first rule of engineering drawing is to standardize the projection that is used. There are many to choose from. This book has dealt with three—isometric, oblique and orthographic. These three are probably the best known. Both isometric and oblique projections have two big disadvantages. Firstly, it is possibie to see only two sides and either the top or the bottom in any one view. It is, of course, possible to draw more than one view, but this brings us to the second disadvantage. On any object, except the simplest, there are curves or arcs or circles. We have seen in Chapters 3 and 6 that, although it is a fairly simple operation to draw these circles, it takes a considerable amount of time and, in industry, time costs money. For these reasons, isometric, oblique and like projections are not used as the standard projection.

The standard projection used is orthographic projection. This is the obvious projection to use because of its many advantages. It presents a true picture of each face: circles remain as circles; ellipses remain as ellipses; horizontal lines remain horizontal; and vertical lines remain vertical. There is no limit to the number of views that you can draw: if the object that you wish to draw is complicated, it is possible to show half a dozen views; if it is simple, two will suffice. Equally important is the fact that, however many views are drawn, they are all related to each other in position.

We still need to decide whether to draw in first or third angle orthographic projection but unfortunately it is impossible to give a definite ruling on this. It is traditional for the British Isles and the Commonwealth to draw in 1st angle projection, but the United States of America and, more recently, the Continental countries have adopted 3rd angle projection. There is no doubt that eventually 3rd angle will become the international standard, but it will take a considerable time. First angle projection is still widely taught in this country, but an examination candidate will need to be familiar with both projections since he may have to answer questions in either.

For full details of both 1st and 3rd angle projections turn back to Chapter 10.

Sections

Sections have already been discussed at some length in Chapter 10 where their main application was in finding the true shape across a body. When sections are used in engineering drawing, although the true shape is still found, the section is really used to show what is inside the object.

A drawing must be absolutely clear when it leaves the drawing board. The person or persons using the drawing to make the object must have all the information that they need presented clearly and concisely so that they are not confused—even over the smallest point.

Suppose that you had to draw the assembly of the three speed gearbox on the rear hub of a pedal cycle. You probably know nothing about the interior of that hub. The reason that you know nothing about it is that you cannot see inside it. If you are to produce a drawing that can be read and understood by anybody, you can draw as many views of the outside as you wish, but your drawing will still tell nothing about the gear train inside. What is really needed is a view of the inside of the hub and this is precisely what a section allows you to show.

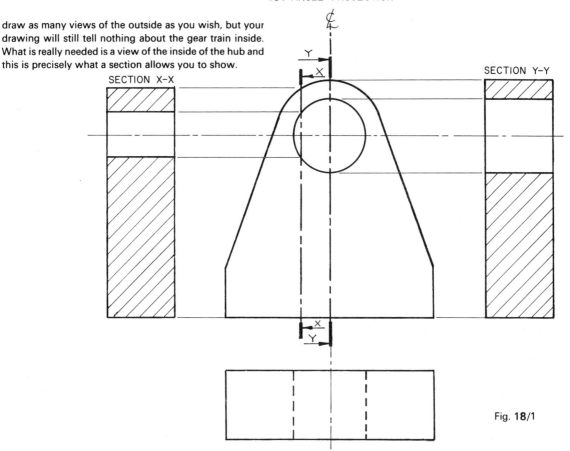

SECTION X–X

SECTION Y–Y

Fig. 18/1

Fig. 18/1 shows two sections projected from a simple bracket. You will notice that the sections are both projected from the Front Elevation. Sections can be projected from any elevation—you are not limited to the front elevation only. Thus, you can project a sectioned F.E. from either the Plan or the End Elevation. A sectioned E.E. is projected from the F.E. and a sectioned Plan is projected from the F.E. It is not usual to project a sectioned E.E. from the Plan nor vice-versa.

The lines X–X and Y–Y are called the sectional cutting planes and this is a good description because you are, in fact, pretending to cut the bracket right through along these lines. Both the sectioned End Elevations are what you would see if you had physically cut the brackets along X–X and Y–Y, removed the material *behind* the cutting planes (that is, the side away from the arrowheads), and projected a normal E.E. with the material removed. To avoid any misinterpretation, and to show the section quite clearly, wherever the cutting plane has cut through material the drawing is hatched. The standard hatching for sectioning is at 45° although it will be seen later that in exceptional circumstances this rule may be broken. You should also note that the cutting plane passes through the hole in the bracket and this is not

hatched. Hatching should only be done when the cutting plane passes through a solid material.

The lines X–X and Y–Y are of a particular nature. They are chain dotted lines thickened at the end of the chain dotted line. The letters X–X and Y–Y are not a random choice either. Sectional cutting planes are usually given letters from the end of the alphabet although you will sometimes see other letters used.

Sectioning is a process which should be used only to simplify or clarify a drawing. You should certainly not put a section on every drawing that you do. There are some engineering details that, if sectioned, lose their identity or create a wrong impression and these items are never shown sectioned. A list of these items is shown below.

Nuts and bolts	Ball bearings and ball races
Studs	Roller bearings and roller races
Screws	Keys
Shafts	Pins
Webs	Gear Teeth

Webs are not shown sectioned because section lines across a web give an impression of solidarity.

195

The question of clarity arises again when considering an assembly, i.e. more than one part. If any of the parts are in the above list, they are not hatched, but a finished product may be composed of several different parts made with several different materials. In the days when productivity was not quite so vital, the draughtsman was a man who turned out drawings that were almost works of art. Since there was no printing as there is to-day only one drawing was made. Each different material was coloured when sectioned and each colour represented a different and specific metal. Later, when drawings were duplicated, colours were no longer used to any great extent and each metal was given its own type of shading and it was still possible to identify materials from the sectioned views. There are now so many types of materials and their alloys in use that it has become impossible to give them all their own type of line, and you can please yourself when deciding which line you will use for a particular section.

There are occasions when hatching at other than 45° is allowable. This is when the hatching lines would be parallel or nearly parallel to one of the sides. Two examples of these are shown in Fig. 18/2.

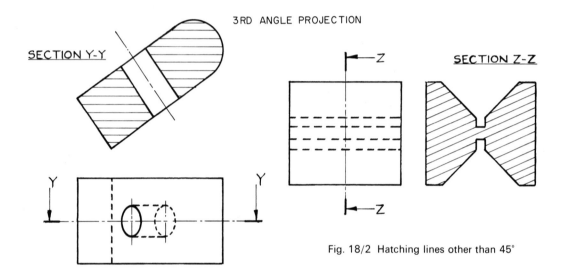

3RD ANGLE PROJECTION

SECTION Y-Y

SECTION Z-Z

Fig. 18/2 Hatching lines other than 45°

If a very large piece of material has to be shown in section, then, in order to save time, it is necessary to hatch only the edges of the piece. An example of this is shown in Fig. 18/3.

If the object that you are drawing is symmetrical and nothing is to be gained by showing it all in section, then it is necessary to show only as much section as the drawing requires. This usually means drawing a half-section.

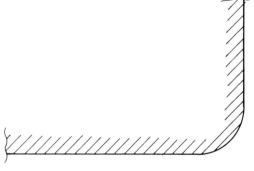

Fig. 18/3 Hatching lines for large bodies

Remembering that sections are used only to clarify a drawing, it is quite likely that you will come across a case where only a very small part of the drawing needs to be sectioned to clarify a point. In this case a part or scrap section is permitted. Two examples of this are shown in Fig. 18/4.

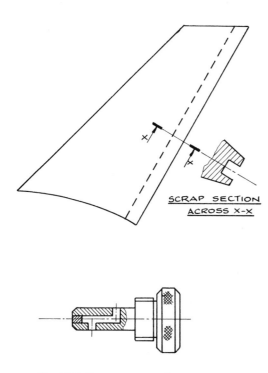

SCRAP SECTION
ACROSS X-X

Fig. 18/4 Part or scrap sections

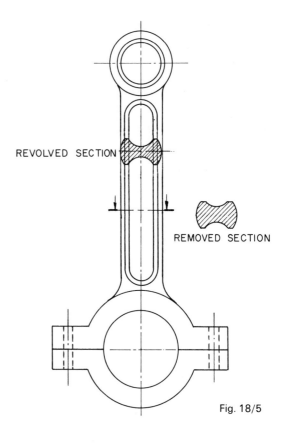

REVOLVED SECTION

REMOVED SECTION

Fig. 18/5

It is often necessary to show a small section showing the true shape across an object. There are two ways of doing this and they are both shown in Fig. 18/5. The revolved section is obtained by revolving the section in its position and breaking the outline to accommodate the section. The removed section is self-explanatory and should be used in preference to the revolved section if there is room on the drawing. It is so very much neater.

When very thin materials have to be shown in section and there is no room for hatching, then they are shown solid. The most common occurrence, of course, is when drawing sheet metal. If two or more parts are shown adjacent, a small place should be left between them, Fig. 18/6.

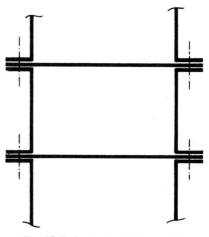

Fig. 18/6 Sectioning thin materials

Fig. 18/8 attempts to show, in a completely hypo-thetical arrangement, as many details as possible about sectioning on one drawing. The drawing shows two lengths of shaft joined by keyed flanges which are bolted together. The shafts are supported by two identical plummer blocks which are, in turn, bolted to the base via four studs for each block. Only one stud is shown on the drawing. This is common practice on a complicated drawing when, rather than simplifying the drawing, it is complicated by too much detail. Note how the section line is thickened where it changes direction, as shown in Fig. 18/7.

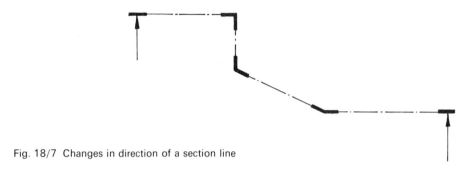

Fig. 18/7 Changes in direction of a section line

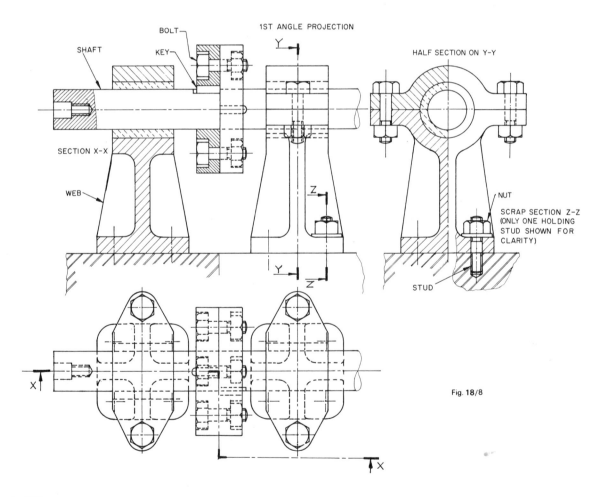

Fig. 18/8

198

SCREW THREADS

The screw thread is probably the most important single component in engineering. The application of the screw thread to nuts, bolts, studs, screws, etc. provides us with the ability to join two or more pieces of material together securely, easily and, most important of all, not permanently. There are other methods of joining materials together but the most widely used ones—riveting, welding and (very common these days) using adhesives, are all permanent. It is true that these methods are cheaper, but when we know that we might have to take the thing apart again we use the screw thread. Since the screw thread is so important it is well worth while looking at the whole subject more closely.

The standard thread, in this country, for many years was the Whitworth; this thread was introduced by Sir Joseph Whitworth in the 1840's. It was the first standard thread; previously a nut and bolt were made together and would fit another nut or bolt only by coincidence. At the time, it was a revolutionary step forward.

The BSW thread and its counterpart the BS Fine thread were the standard threads in this country until metrication and will probably be in use for many years.

However, the U.S.A. developed and adopted the Unified thread as their standard and countries using the Metric System of measurement had their own Metric thread forms. It became increasingly obvious that an international screw thread was needed.

As far as this country was concerned, the breakthrough came when it was decided that British Industry should adopt the Metric System of Weights and Measures. The International Organisation for Standardization (ISO) has formulated a complex set of standards to cover the whole range of engineering components.

Their thread, the ISO is now the international standard thread. The ISO and Unified thread profiles are identical. The Unified thread is the Standard International thread for countries which are still using Imperial units.

The ISO basic thread form is shown in Fig. 18/9.

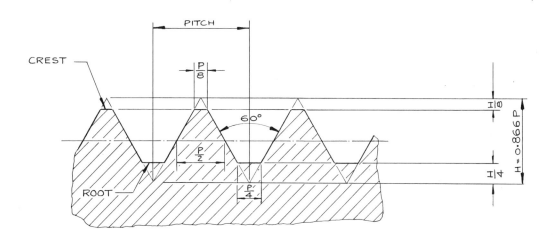

Fig. 18/9 Basic form of ISO thread

You will notice that the thread is thicker at the root than at the crest. This is because the stresses on the thread are greater at the root and the thread needs to be thicker there if it is to be stronger.

In practice, since there is nothing gained by having the root and crest of a nut and bolt in contact, and because 'square' corners are difficult to manufacture, the ISO thread form is usually modified to that shown in Fig. 18/10

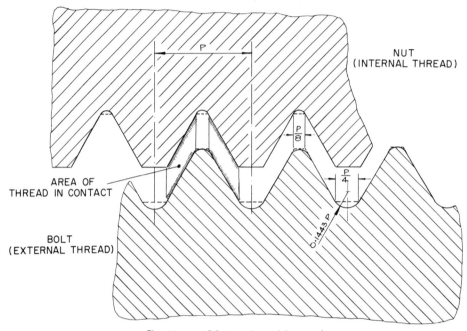

Fig. 18/10. ISO thread used in practice

You can see that the contact will be only on the flanks.

There is more than one type of ISO thread. There are 12 series of threads ranging from the widely used coarse thread series, which is used on bolts from 1·6 mm dia. to 68 mm dia., to a 6 mm constant pitch series with sizes from 70 mm dia. to 300 mm dia. The whole range of thread series has the same basic profile and full details can be found in BS 3643. The fine thread series (the

equivalent to the redundant BSF) ranges from 1·8 mm to 68 mm.

The British Standard Whitworth thread has now been superseded by the ISO thread. However, the International Organisation for Standardization has adopted the Whitworth profile for pipe threads. It is called the British Standard Pipe Thread, Fig. 18/11.

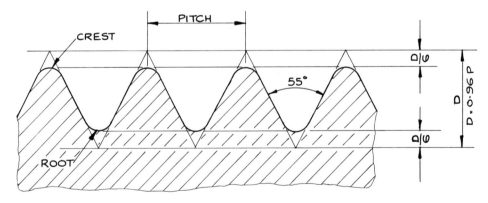

Fig. 18/11 British standard pipe thread
(British standard Whitworth thread profile)

There are some special threads that have been designed to fulfil functions for which a vee thread would be inadequate. Some are shown.

The square thread, Fig. 18/12, is now rarely used because it has been superseded by the Acme thread. Its main application is for transmitting power since there is less friction than with a vee thread.

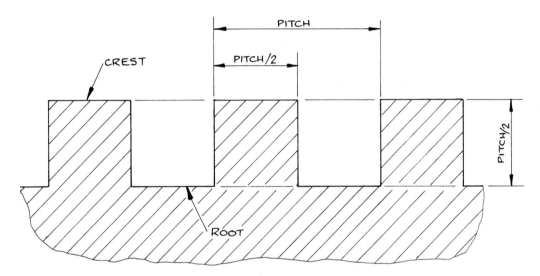

Fig. 18/12 Square thread profile

The Acme thread, Fig. 18/13, is extensively used for transmitting power. The thread form is easier to cut than the square thread because of its taper and, for the same reason, it is used on the lead screw of lathes where the half-nut engages easily on the tapered teeth.

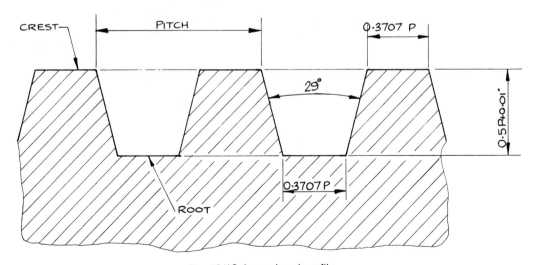

Fig. 18/13 Acme thread profile

The Buttress thread, Fig. 18/14, combines the vee thread and the square thread without retaining any of their disadvantages. It is a strong thread and has less friction than a vee thread. Its main application is on the engineer's vice although it is sometimes seen transmitting power on machines.

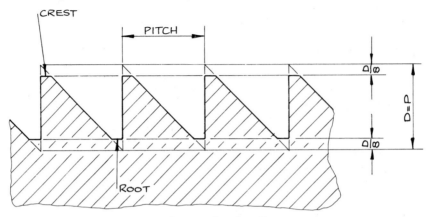

Fig. 18/14 Buttress thread profile

Drawing screw threads

Drawing a screw thread properly is a long and tedious business. A square thread has been drawn in full in Fig. 15/12 and you can see that this type of construction would take much too long a time on a drawing that has several threads on it and would be physically impossible on a small thread.

There are conventions for drawing threads which make life very much easier. Three conventional methods of representing screw threads are shown in Fig. 18/15. The

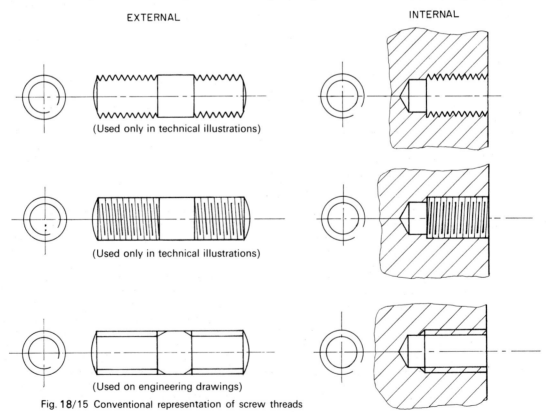

Fig. 18/15 Conventional representation of screw threads

top two methods are not used on engineering drawings any more. One of these illustrates the shape of the thread and the other has lines representing the thread crest and root. The bottom drawing shows how to draw a screw thread on an engineering drawing; the parallel lines represent the thread crest and root.

The only convention which shows whether the thread is right- or left-handed is the second one. This is not much of an advantage because the thread has to be dimensioned and it is a simple matter to state whether a thread is right- or left-handed. Left-hand threads are rarely met with and, unless specifically stated, a thread is assumed to be right-handed.

Fig. 18/15 shows the convention for both external and internal threads. It should be explained that, on the drawings for internal threads, the thread does not reach to the bottom of the hole. When an internal thread is cut, the material is first drilled a little deeper than is actually required. The diameter of the hole is the same as the root diameter of the thread and is called the tapping diameter. The screw thread is then cut with a tap, but the tap cannot reach right to the bottom of the hole and some of the tapping hole is left. The cutting angle of the drill, for normal purposes, is 118°—almost 120°. Thus, a 60° set square is used to draw the interior end of an internal screw thread.

Nuts and Bolts

The most widespread application of the screw thread is the nut and bolt. Whenever a nut or bolt is drawn, it is essential that the first view drawn is the one which shows the regular hexagon. If the across flats (A/F) dimension is given, draw a circle with that diameter. If it is not, look in Appendix A under the appropriate thread size. Construct a regular hexagon round the circle with a 60° set square. Project the corners of the hexagon onto the side view of the nut and bolt and mark off the thickness of the nut or bolt head (see Appendix A).

Nuts and bolts are chamfered and, when viewed from the side, this chamfer is seen as radii on the sides of the nut or bolt. If you ensure that the first view projected from the hexagon is the one which shows three faces of the nut (the other view shows only two faces) you can draw a radius equal to D, the diameter of the thread, on the centre flat. The intersection of this radius and the corners of the neighbouring flats determines the size of the two smaller radii. These must start at this intersection, finish at the same height on the next corner, and touch the top of the nut or bolt at the centre of the flat. This may be done by trial and error with compasses, or with radius curves. Remember that the centre of the radius lies midway between the sides. This view is completed by drawing the 30° chamfer which produced the radii.

The third view of the nut or bolt is drawn in a similar fashion. The width and heights are projected from the two existing views and the radii are found in the same way as shown on the other view.

Two types of standard nuts and bolts are shown in Fig. 18/16. Type A is standard. Type B has a 'washer face' underneath the head of the bolt and on one face of the nut.

The threaded end of the bolt may be finished off with a spherical radius equal to $1\frac{1}{2}D$ or a 45° chamfer to just below the root of the thread. Both of these enable the nut to engage easily and leave no sharp projections. The thread on the nut is also chamfered to assist easy engagement.

The length of a bolt is determined simply by the use to which the bolt is to be put. There is a very large selection of bolt lengths for all diameters. The bolt should not protrude very far past the nut and so there is no need to thread all of the shank. The amount of thread on a bolt is given in the table below.

LENGTH OF BOLT	LENGTH OF THREAD
Up to and including 125 mm	2d + 6 mm
Over 125 mm and up to and including 200 mm	2d + 12 mm
Over 200 mm	2d + 25 mm

d is the diameter of the bolt

These lengths are the minimum thread lengths.

Bolts which are too short for minimum thread lengths are called screws.

An ISO metric nut or bolt is easily recognised by the letter 'M' or ISO M on the head or flats.

Designation of ISO Screw Threads

The Coarse Series ISO thread is only one of twelve different threads in the ISO series. This thread, like the Fine Thread Series, has a pitch which varies with the diameter of the bolt. The remaining ten thread series have constant pitches, whatever the diameter of the thread.

All the series except the Coarse Thread Series are used in special circumstances. The vast majority of threads used come from the Coarse Thread series.

The method used on drawings for stating an ISO thread is quite simple. Instead of stating the thread form and series you need only use the letter 'M'. The diameter of the thread is stated immediately after 'M'. Thus M12 is ISO thread form, 12 mm dia. thread and M20 is ISO thread form, 20 mm dia. thread. In many countries the designation shown above is used to denote a Coarse Series Thread. If a thread is used from a constant pitch series, it is added after, so that M14 × 1.5 is a 14 mm diameter ISO thread with a constant pitch of 1.5 mm.

However, in this country the British Standard requires that the pitch be included in the Coarse Series Threads. Thus, a thread with the designation M30 × 3.5 is a Coarse Series ISO thread with a pitch of 3.5 mm.

A thread with the designation M16 × 2 is a Coarse Series ISO thread with a pitch of 2 mm.

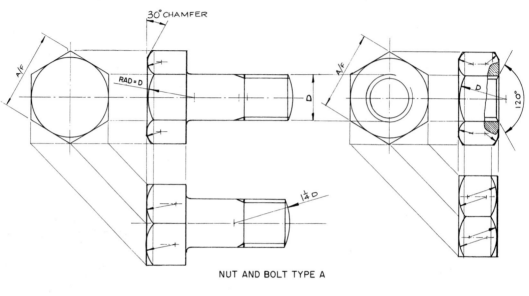

NUT AND BOLT TYPE A

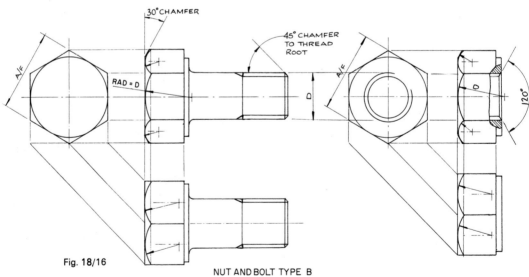

Fig. 18/16

NUT AND BOLT TYPE B

SEE APPENDIX A FOR NUT, BOLT AND WASHER PROPORTIONS

There are further designations concerned with the tolerances, or accuracy of manufacture, but these are beyond the scope of this book.

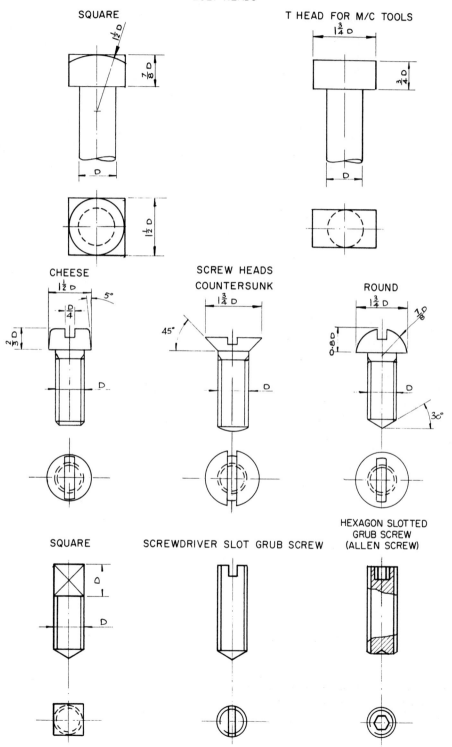

Fig. 18/17 Bolt heads and screw heads

Types of Bolts and Screws

There are many types of heads for bolts and screws apart from the standard hexagonal head. Some are shown in Fig. 18/17.

Fig. 18/17 shows only a few types of bolt and screw heads that are in use. There are Wedged-Shaped Heads, Tommy Heads, Conical Heads, Hook Bolts and Eye Bolts. There are Small, Medium and Large Headed Square screws, 60°, 120° and 140° countersunk screw heads with straight slots, cross slots and hexagonal slots. There are Instrument Screws and Oval Cheese-Headed Screws to name only a few. The dimensions for all these screws can be obtained from any good engineering handbook.

DIMENSIONING

When an engineering drawing is made, dimensioning is of vital importance. All the dimensions necessary to make the articles drawn must be on the drawing and they must be presented so that they can be easily read, easily found, and not open to misinterpretation. A neat drawing can be spoilt by bad dimensioning.

In British drawing practice the decimal point is shown in the usual way, i.e. 15.26. On the Continent, however, the decimal point in Metric units is a comma, i.e. 15,26 or 0,003.

Also, in the Metric system a space is left between every three digits, i.e. 12 056.0 or 0.002 03. Note that values less than unity are prefixed by a nought. Engineering drawings are usually dimensioned in millimetres, irrespective of the size of the dimension, but the centimetre and metre are also sometimes used.

There are many rules about how to dimension a drawing properly, but it is unlikely that two people will dimension the same drawing in exactly the same way. However, remember when dimensioning that you must be particularly neat and concise, thorough and consistent. The following rules must be adhered to when dimensioning.

1. Projection lines should be thin lines and should extend from about 1 mm from the outline to 3 mm to 6 mm past the dimension line.
2. The dimension line should be a thin line and terminate with arrowheads at least 3 mm long and these arrowheads must touch the projection lines.
3. The dimension may be inserted within a break in the dimension line or be placed on top of the dimension line.
4. The dimensions should be placed so that they are read from the bottom of the paper or from the right-hand side of the paper.
5. Dimension lines should be drawn *outside* the outline, whenever possible, and should be kept well clear of the outline.
6. Overall dimensions should be placed outside the intermediate dimensions.

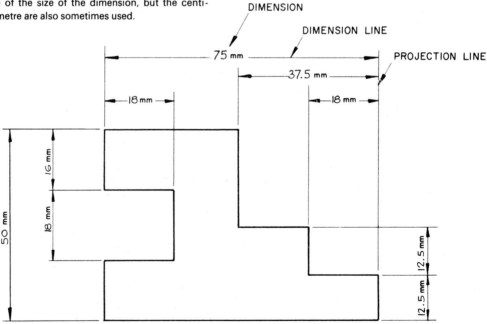

Fig. 18/18 Illustrating rules 1—6

7. Centre lines must *never* be used as dimension lines, They may be used as projection lines.
8. Diameters may be dimensioned in one of two ways. Either dimension directly across the circle (*not* on a centre line), or project the diameter to outside the outline. 'Diameter' is denoted by the symbol ϕ placed in front of the dimension.
9. When dimensioning a radius, you must, if possible, show the centre of the radius. The actual dimension for the radius may be shown either side of the outline but should, of course, be kept outside if possible. The word radius must be abbreviated to R and placed in front of the dimension.
10. When a diameter or a radius is too small to be dimensioned by any of the above methods, a leader may be used. The leader line should be a thin line and should terminate on the detail that it is pointing to with an arrowhead or, within an outline, with a dot. Long leader lines should be avoided even if it means inserting another dimension. The leader line should always meet another line at an acute angle.
11. Dimensions should *not* be repeated on a drawing. It is necessary to put a dimension on only once, however many views are drawn. There is one exception to this rule. If, by inserting one dimension, it saves adding up lots of small dimensions then this is allowed. These types of dimensions are called auxiliary dimensions and are shown to be so either by underlining the dimensions or putting it in brackets.

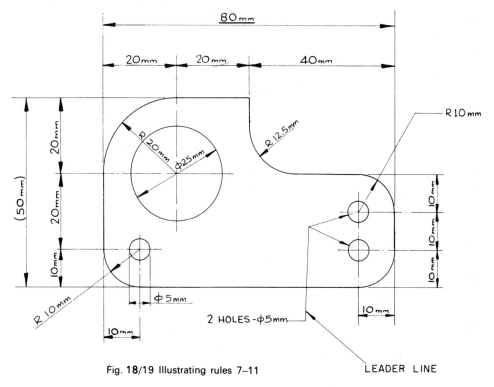

Fig. 18/19 Illustrating rules 7–11

207

12. Unless unavoidable, do *not* dimension hidden detail. It is usually possible to dimension the same detail on another view.

13. When dimensioning angles, draw the dimension lines with a compass; the point of the compass should be on the point of the angle. The arrowheads may be drawn either side of the dimension lines, and the dimension may be inserted between the dimension lines or outside them. Whatever the angle, the dimension must be placed so that it can be read from either the bottom of the paper or from the right-hand side.

14. If a lot of parallel dimensions are given, it avoids confusion if the dimensions are staggered so that they are all easier to read.

15. If a lot of dimensions are to be shown from one projection line (often referred to as a *datum line*), either of the methods shown in Fig. 18/20 may be used. Note that in both methods, the actual dimension is close to the arrowhead and not at the centre of the dimension line.

16. If the majority of dimensions on a drawing are in one unit, it is not necessary to put on the abbreviation for the units used, i.e. cm or mm. In this case, the following note must be printed on your drawing.

UNLESS OTHERWISE STATED, DIMENSIONS
ARE IN MILLIMETRES

3RD ANGLE PROJECTION

UNLESS OTHERWISE STATED, ALL DIMENSIONS ARE IN MILLIMETRES

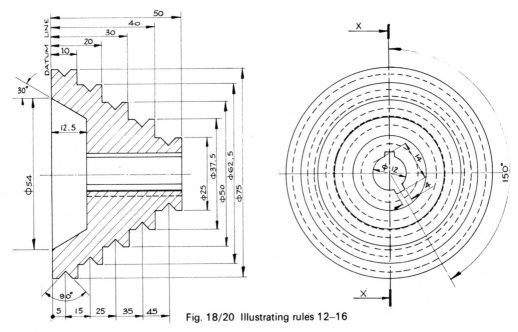

Fig. 18/20 Illustrating rules 12–16

208

17. If a very large radius is drawn, whose centre is off the drawing, the dimension line is drawn with a single zig-zag in it.
18. Dimensioning small spaces raises its own problems and solutions. Some examples are shown in Fig. 18/21

There are one or two more rules that do not require illustrating.
19. If the drawing is to scale, the dimensions put on the drawing are the actual dimensions of the component and not the size of the line on your drawing.

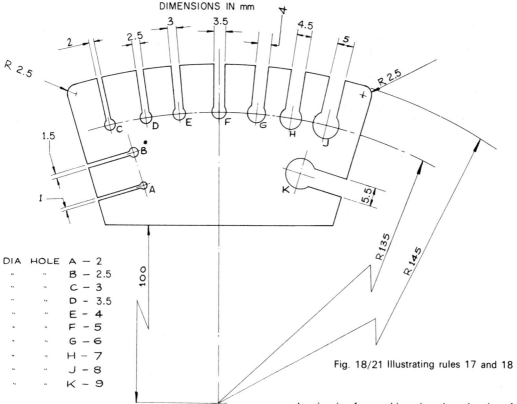

DIMENSIONS IN mm

DIA HOLE A – 2
" " B – 2.5
" " C – 3
" " D – 3.5
" " E – 4
" " F – 5
" " G – 6
" " H – 7
" " J – 8
" " K – 9

Fig. 18/21 Illustrating rules 17 and 18

The above nineteen rules do not cover all aspects of dimensioning (there are a whole new set on toleranced dimensions alone) but they should cover all that is necessary up to 'O' level G.C.E. Dimensioning properly is a matter of applying common sense to the rules because no two different drawings can ever raise exactly the same problems. Each drawing that you do needs to be studied very carefully before you begin to dimension it.

Examination questions often ask for only five or six 'important' dimensions to be inserted on the finished drawing. The overall dimensions—length, breadth and width—are obviously important but the remaining two or three are not so obvious. The component or assembled components need to be studied in order to ascertain the function of the object. If, for instance, the drawing is of a bearing, then the size of the bearing is vitally important because something has to fit into that bearing. If the drawing is of a machine vice, then the size of the vice jaws should be dimensioned so that the limitations of the vice are immediately apparent. These are the types of dimensions that should make up the total required.

CONVENTIONAL REPRESENTATIONS

There are many common engineering details that are difficult and tedius to draw. The screw thread is an example of this type of detail and it has been shown earlier in this part of the book that there are conventional ways of drawing screw threads which are very much simpler than drawing out helical screw threads in full.

Fig. 18/22 shows some more engineering details and alongside the detailed drawing is shown the conventional representation for that detail. These conventions are designed to save time and should be used wherever and whenever possible.

These are not all the standard conventions but the rest are beyond the scope of this book. The interested student can find the rest in BS308.

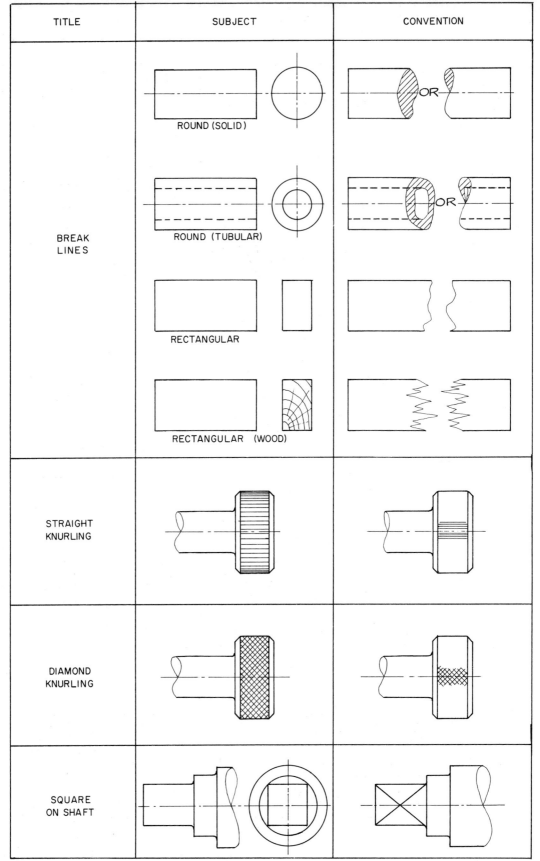

TITLE	SUBJECT	CONVENTION
BREAK LINES	ROUND (SOLID)	OR
	ROUND (TUBULAR)	OR
	RECTANGULAR	
	RECTANGULAR (WOOD)	
STRAIGHT KNURLING		
DIAMOND KNURLING		
SQUARE ON SHAFT		

Fig. 18/22 (a)

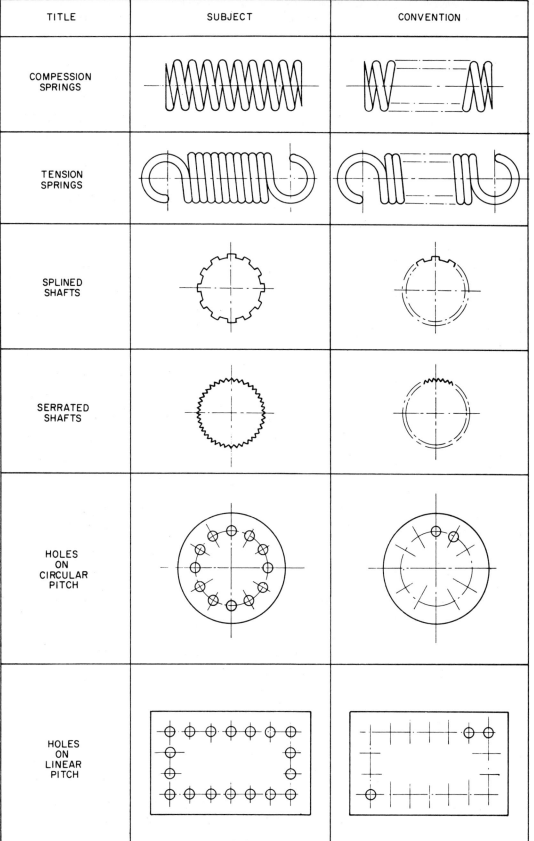

TITLE	SUBJECT	CONVENTION
COMPESSION SPRINGS		
TENSION SPRINGS		
SPLINED SHAFTS		
SERRATED SHAFTS		
HOLES ON CIRCULAR PITCH		
HOLES ON LINEAR PITCH		

211

Fig. 18/22 (b)

TITLE	SUBJECT	CONVENTION
TREATMENT OF SYMMETRICAL PARTS	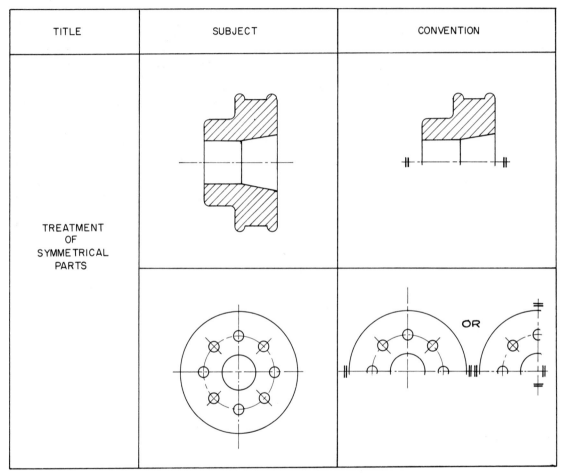	

Fig. 18/22 (c)

(Note the short thick double line at the end of each centre line.)

Machining Symbols

The shape of an engineering component can be determined in several ways. The component may be forged, cast, drawn, etc. After one or more of these processes, it is quite likely that some machining will have to be done. It is therefore important that these machined faces be indicated on the drawing. The method recommended by BS 308 is shown in Fig. 18/23 but this is not the only method in use. Sometimes, the letter 'f' is written over the face to be machined. This letter 'f' stands for 'finish'.

The small tick shows only that that particular face has to be machined. It does not show how it is to be machined, nor does it show how smooth the finish is to be. The method of machining—turning, milling, grinding, etc. —is not normally put on a drawing but the standard of finish is.

Surface roughness

This short section on roughness symbols is beyond the G.C.E. 'O' level syllabus, but it is well worth looking at.

The standard of finish, or roughness of a surface, is of vital importance in engineering. The degree of roughness permitted depends on the function of the component. When two pieces of metal slide against each other, as in the case of a bearing, the finish on both parts must be very smooth or the bearing will overheat and 'seize'. On the other hand, smooth finishes are expensive to produce and should be kept to a sensible minimum. The moving parts of an internal combustion engine can be so well finished that it is not necessary to 'run in' the car but this is an expensive process applied only on very expensive motor cars. The mass-produced car needs several hundred miles of careful driving while the surfaces 'wear smooth'.

If the surface of a piece of machined metal is magnified it will look like a range of very craggy mountains. The surface roughness is the distance from the highest 'peak' to the lowest 'valley'. This roughness is measured in micro-metres and one micro-metre is one millionth part of a metre. Not only can a surface be made smooth to one micro-metre but it can also be measured to one micro-metre.

The British Standard index numbers of surface roughness are 0.025: 0.05: 0.1: 0.2: 0.4: 0.8: 1.6: 3.2: 6.3: 12.5 and 25.0. A surface roughness of from 0.025 to 0.2 can be obtained by lapping or honing. 0.4 can be obtained by grinding and 0.8 by careful turning, rough grinding, etc. The surface roughness number is shown within the vee of the machining symbol. A tolerance on surface roughness is shown as a fraction, with the maximum

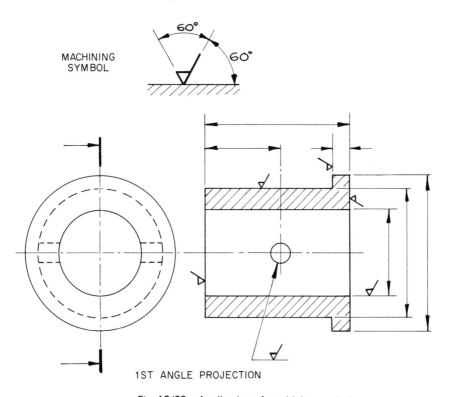

1ST ANGLE PROJECTION

Fig. 18/23 Application of machining symbol

roughness number above the minimum roughness number, Fig. 18/24.

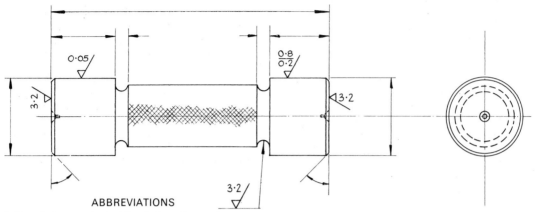

ABBREVIATIONS

A list of the standard abbreviations for some general engineering terms is shown below. These abbreviations are used to save time, and space on drawings.

Although the abbreviations are shown in block capital letters, small letters may be used. For unit abbreviations always use small letters. The abbreviations are the same in the singular and the plural. Note that full stops are not used except when the abbreviation makes another word, e.g. Number becomes No. and Figure becomes Fig.

These are by no means all the standard abbreviations. They should contain all that are required at this stage but should the student require the full list of abbreviations, he will find them in BS 1991.

Fig. 18/24 Application of surface roughness symbols

Term	Abbreviation
Across flats	A/F
British Standard	BS
Centimetre	cm
Centre line	CL or ₵
Chamfered	CHAM
Cheese head	CH HD
Countersunk	CSK
Counterbore	C'BORE
Degree (of angle)	°
Diameter (preceding a dimension)	ϕ
Drawing	DRG
Figure	FIG.
First Angle Projection	
Hexagon	HEX
Internal diameter	I/D
Kilogram	kg
Left hand	LH
Machine	M/C
Machined	M/CD
Machinery	M/CY
Material	MATL
Maximum	MAX

Term	Abbreviation
Metre	m
Millimetre	mm
Minimum	MIN
Minute (of angle)	'
Number	NO.
Outside diameter	O/D
Per	/
Pitch circle diameter	PCD
Radius (preceding a dimension)	R
Right hand	RH
Round head	RD HD
Second (of angle)	"
Square (in a note)	SQ
Square (preceding a dimension)	□
Square metre	m²
Standard	STD
Third Angle Projection	

SCREW THREADS

Term	Abbreviation
British Standard Pipe	BSP
International Organization for Standardization	ISO
Système International	SI
Threads per inch	TPI
Undercut	U'CUT
Unified Coarse	UNC
Unified Fine	UNF
Unified Selected	UNS

FRAMING AND TITLE BLOCKS

Drawing paper comes in standard sizes. The largest size found in schools and colleges is usually size A1, 841 mm

× 594 mm. The average student will do the majority of his drawing on size A2, 594 mm × 420 mm and this is also the size that most examination questions are answered on.

If it is possible, an engineering drawing should be so positioned that it makes the maximum use of the available space. The positions of the elevations to be drawn must be calculated before the drawing is started. The calculations are simple enough and are dependent upon the overall size of the component.

Assuming that these sizes are A, B and C for the maximum length, breadth and height respectively, and assuming that the spaces between the three elevations to be drawn and the edge of the paper are to be equal, a specimen layout is shown in Fig. 18/25.

It is not necessary to use exact figures for dimensions A, B and C. They should be approximated so that the calculations are simplified.

If the size of the component that is drawn is such that the drawn views fit the paper neatly without large gaps between the elevations, then the frame around the drawing should be at least 15 mm from the edge of the paper all the way round

The distance between the elevations should not be larger than would be required to fully dimension the drawing neatly. If the paper is obviously much larger than is necessary, and this often happens in examinations, do not attempt to fill the paper and thus have large spaces between the drawn views. Position the elevations so that they are not too far apart and draw the frame round the

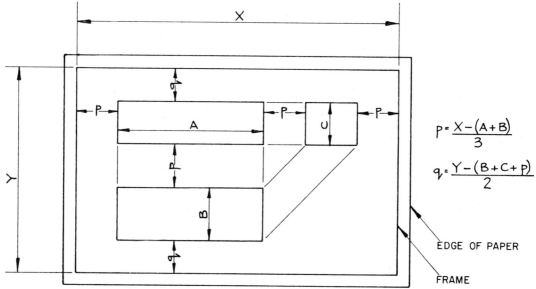

$$P = \frac{X - (A + B)}{3}$$

$$q = \frac{Y - (B + C + P)}{2}$$

Fig. 18/25 Positioning of views to be drawn

drawing with no regard to the edge of the paper. The frame should neatly encompass the drawn views, and more space must be left at the bottom (for the title blocks) than is left at the sides and the top.

On an industrial drawing, there is a lot of additional information to add to a drawing after the actual drawing is finished. Apart from the drawing number and title, the name of the firm, the scale, the date, a materials or parts list, the job or order number, any treatments or finishes, a key to machining and other symbols, tolerances not particularly mentioned, tool and gauge references, general notes and quantities and cross references to other associated drawings are among the details that must be inserted on the drawing. Boxes or title blocks are usually already on the paper that the draughtsman uses, and the above information is inserted in the appropriate box.

In normal classwork, the student need only put his name and form but more information is usually required by examiners and it is good practice to insert on a finished

drawing the following information:
    Name
    School or college
    Title
    Scale
    System of projection
    Date

The above information must be placed in title blocks. These blocks are drawn in one of two positions. They are either in a group in the bottom right-hand corner of the drawing or they are spread out along the bottom of the drawing. In exceptional circumstances, e.g. when there is plenty of room at the top of the drawing but none at the bottom, the blocks may be positioned elsewhere, but in normal circumstances they are in one of the two positions shown in Fig. 18/26.

215

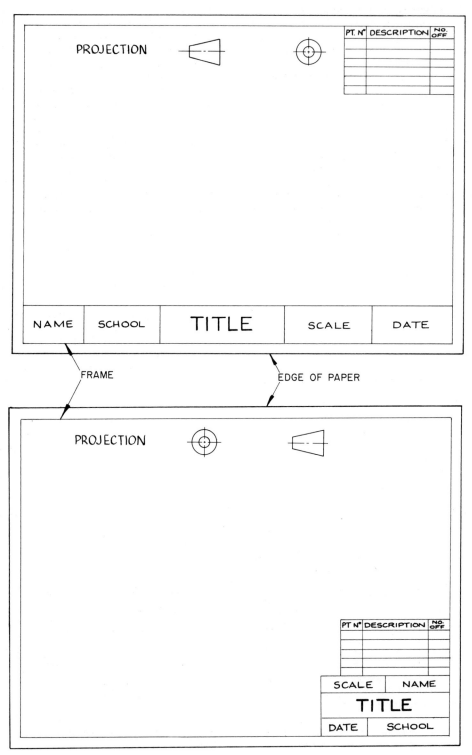

Fig. **18**/26 Layout of title boxes

Note that the title is given prominence over all the other information. Also note that the system of projection is at the top of the paper. This is because it is really a note on the drawing rather than an additional piece of information. The list of part numbers would be used only if several parts were drawn on the same drawing and would not, therefore, be shown on every drawing.

ASSEMBLY DRAWINGS

There are not many engineering items that are completely functional by themselves. There are some, a spanner or a rule for instance, but even a simple object like a wood chisel has three components and a good pair of compasses may have twelve component parts. Each part should be drawn and dimensioned separately and then a drawing is made of all the component parts put together. This is called an assembly drawing. The student at school or college is often instructed to draw the assembled components only and is shown the dimensioned details in no particular order. If the assembly is particularly difficult, the parts are often shown in an exploded view and the assembly presents no difficulty. The assembled parts may form an object which is easily recognizable, but the real problem occurs when there seems to be no possible connection between any of the component parts. In an examination, when loss of time must be avoided at all costs, the order of assembly needs to be worked out quickly.

The only approach is to view the assembly somewhat like a jig-saw puzzle. The parts must fit together and be held together, either because they interlock or because there is something holding them together.

Learn to look for similar details on separate components. If there is an internal square thread on one component and an external thread of the same diameter on another component, the odds are that one screws inside the other. If two different components have two or more holes with the same pitch, it is likely that they are joined at those two holes. A screw with an M10 thread must fit an M10 threaded hole. A tapered component must fit another tapered component.

The important thing, particularly in examinations, is to start drawing. Never spend too long trying to puzzle out an assembly. There is always an obvious component to start drawing, and, while you are drawing that, the rest of the assembly will become apparent as you become more familiar with the details.

SOME ENGINEERING FASTENINGS

We have already seen how to draw a standard nut and bolt. Some other types of bolts and screws are shown in Fig. 18/17 with their principal dimensions. There are, however, many other types of fastenings in everyday use in industry, and some of them are shown below.

The stud and set bolt

The stud and set bolt (sometimes called a tap bolt or cap screw) are used when it is impossible or impractical to use a nut and bolt. Fig. 18/27 shows both in their final positions. They are both screwed into a tapped hole in the bottom piece of material. The top piece of material is drilled slightly larger than the stud or screw and is held in position by a nut and washer in the case of the stud, and by the head of the set bolt and washer in the case of the set bolt.

The stud would be used when the two pieces of material were to be taken apart quite frequently; the set bolt would be used if the fixing was expected to be more permanent.

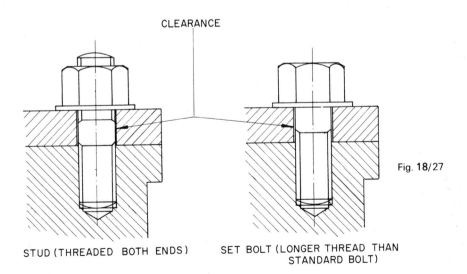

CLEARANCE

Fig. 18/27

STUD (THREADED BOTH ENDS)

SET BOLT (LONGER THREAD THAN STANDARD BOLT)

Locking devices

Constant vibration tends to loosen nuts and, if a nut is expected to be subjected to vibration, a locking device is often employed. There are two basic groups of locking devices: one group increases the friction between the nut and the bolt or stud; the other group is more positive and is used when heavy vibration is anticipated or where the loss of a nut would be catastrophic. Fig. 18/28 shows five common locking devices.

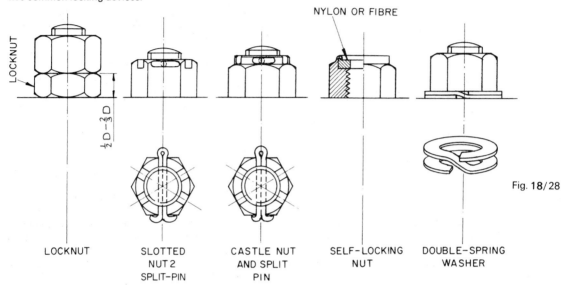

Fig. 18/28

D=DIAMETER OF THREAD

SEE APPENDIX B FOR PROPORTIONS OF SLOTTED AND CASTLE NUTS

The *locknut* is very widely used. The smaller nut should be put on first and, when the larger nut is tightened, the two nuts strain against each other. The smaller nut pushes upwards. The reaction in the larger nut is to push downwards against the smaller nut and, since it must move upwards to undo, it is locked in position.

The *slotted nut* and *castle nut* are used when the nut must not undo. The nut is tightened and then a hole is drilled through the bolt through one of the slots. A split pin is inserted and the ends are bent over. A new split pin should be used each time the nut is removed.

The *self-locking* nut is now very widely used. The nylon or fibre washer is compressed against the bolt thread when the nut is tightened. This nut should only be used once since the nylon or fibre is permanently distorted once used.

The *spring washer* pushes the nut up against the bolt thread, thus increasing the frictional forces. It is the least effective of the locking devices shown and should only be used where small vibrations are expected.

There are many other types of locking device and full descriptions can be found in any good engineering handbook.

Rivets and Riveted Joints

A rivet is used to join two or more pieces of material together permanently. The enormous advances in welding and brazing techniques, and the rapidly increasing use of bonding materials have led to a slight decline in the use of rivets. However, they remain an effective method of joining materials together, and, unlike welding and bonding, require very little special equipment or expensive tools when used on a small scale.

The rivet is usually supplied with one end formed to one of the shapes shown in Fig. 18/29. The other end is hammered over and shaped with a tool called a 'dolly'.

d = RIVET DIAMETER = 1·2t

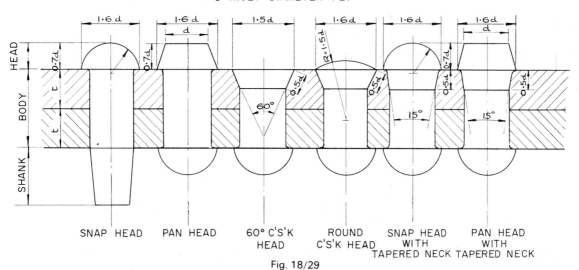

SNAP HEAD   PAN HEAD   60° C'S'K   ROUND   SNAP HEAD   PAN HEAD
                          HEAD   C'S'K HEAD   WITH   WITH
                                              TAPERED NECK   TAPERED NECK

Fig. 18/29

When rivets are used they must be arranged in patterns. The materials to be joined must have holes drilled in them to take the rivets and these holes weaken the material, particularly if they are too close together. If the rivets are placed too close to the edge of the material the joint will be weakened. The two basic joints are called lap and butt joints. Fig. 17/30 shows four examples. There is no limit to the number of rows of rivets, nor to the number in each row, but the spacing, or pitch of the rivets, must be as shown.

There are other types of rivet, the most important group being those used for thin sheet materials. These are beyond the scope of this book and details can be found in any good engineering handbook.

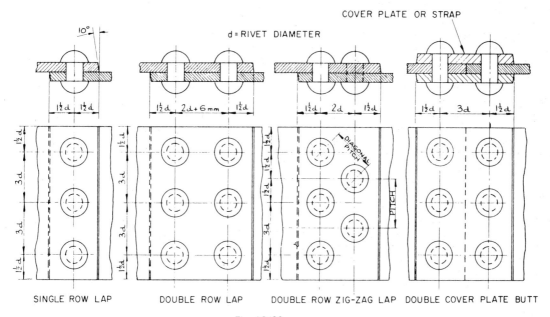

SINGLE ROW LAP   DOUBLE ROW LAP   DOUBLE ROW ZIG-ZAG LAP   DOUBLE COVER PLATE BUTT

Fig. 18/30

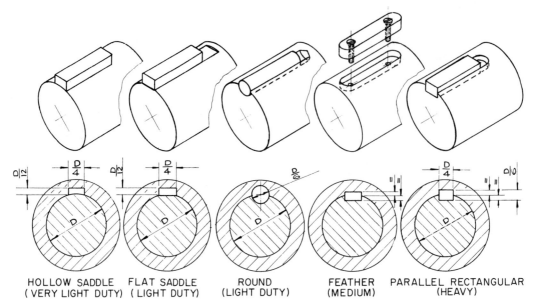

HOLLOW SADDLE (VERY LIGHT DUTY)   FLAT SADDLE (LIGHT DUTY)   ROUND (LIGHT DUTY)   FEATHER (MEDIUM)   PARALLEL RECTANGULAR (HEAVY)

PROPORTIONS ARE APPROXIMATED FOR DRAWING ONLY. FOR EXACT DIMENSIONS SEE ISO TC/16

Fig. 18/31

Keys, Keyways and Splines

A key is a piece of metal inserted between the joint of a shaft or hub to prevent relative rotation between the shaft and the hub. One of the commonest applications is between shafts and pulleys.

There is a wide variety of keys, designed for light and heavy duties, for tapered and parallel shafts and to allow or prevent movement of the hub along the shaft, Figs. 18/31 and 18/32.

Saddle keys are suitable for light duty only since they rely on friction alone.

Round keys are easy to install because the shaft and hub can be drilled together but they are suitable for light duty only.

Feather keys and *parallel keys* are used when it is desired that the hub should slide along the shaft, yet not be allowed to rotate around the shaft.

Taper keys are used to prevent sliding, and the *Gib head* allows the key to be extracted easily.

Woodruff keys are used on tapered shafts. They adjust easily to the taper when assembling the shaft and hub.

The exact dimensions for keys and keyways for any given size of shaft can be found in ISO TC/16.

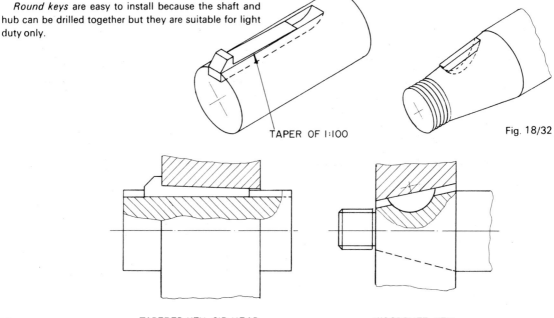

TAPER OF 1:100

Fig. 18/32

TAPERED KEY–GIB HEAD (HEAVY DUTY – EASILY REMOVED)

WOODRUFF KEY (FOR TAPERED SHAFT)

Keyways are machined out with milling machines. If a horizontal milling machine is used, the resulting keyway will look like the one to the left of Fig. 18/33. If a vertical

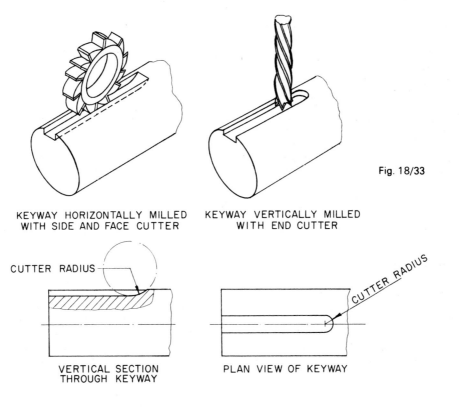

Fig. 18/33

KEYWAY HORIZONTALLY MILLED
WITH SIDE AND FACE CUTTER

KEYWAY VERTICALLY MILLED
WITH END CUTTER

CUTTER RADIUS

VERTICAL SECTION
THROUGH KEYWAY

CUTTER RADIUS

PLAN VIEW OF KEYWAY

milling machine is used, the resulting keyway will look like the one to the right of Fig. 18/33. In both cases, the end of the milled slot has the same profile as the cutter.

If a shaft is carrying very heavy loads, it should be obvious that the load is transferred to the hub (or vice versa) via the key. This means that the power that any shaft or hub can transmit is limited by the strength of the key. If heavy loading is expected, the shaft and hub will be splined, Fig. 18/34. The number of splines will be dependent upon the load to be carried; the greater the number of splines, the greater the permissible loading.

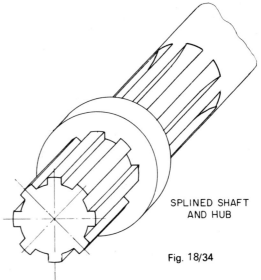

SPLINED SHAFT
AND HUB

Fig. 18/34

Cottered Joints

Keys and splines are used when shafts are subjected to torsional (twisting) loads. If two shafts have to be joined together and then be subjected to tension or compression (push-pull) a different type of fastening is needed. One method is to use a cotter. Two examples of cottered joints are shown in Fig. 18/35. To the left is a connection for two round shafts, and to the right is the connection for two square or rectangular shafts.

The whole assembly locks together as follows:

As the tapered cotter is forced downwards it reacts against faces A and B and tries to draw the shafts together. The smaller shaft cannot be pulled in any further, either because of the collar at C, or because the shafts meet at D. Thus, the more the cotter is forced down, the tighter the assembly. The shafts are easily separated by knocking out the cotter from underneath.

The square or rectangular bar is opened out to form a Y-shaped fork. The smaller bar fits inside the arms of this Y. A gib is used to prevent the arms from spreading when the cotter is hammered in. This problem does not arise on round bars because the larger bar wraps completely round the smaller one.

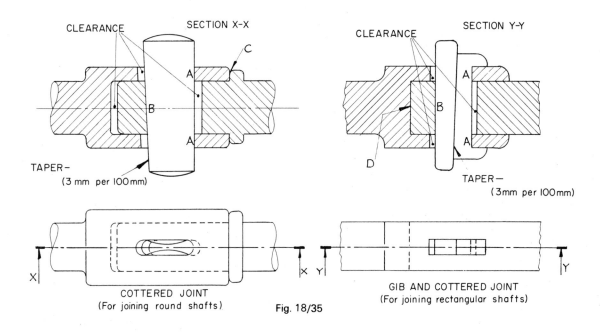

SECTION X-X

CLEARANCE

C

A

B

A

TAPER—
(3 mm per 100 mm)

COTTERED JOINT
(For joining round shafts)

SECTION Y-Y

CLEARANCE

A

B

A

D

TAPER—
(3mm per 100mm)

GIB AND COTTERED JOINT
(For joining rectangular shafts)

Fig. 18/35

Three Worked Examples

Three examples from recent examination papers are shown in the next few pages. The question, as it was set, is shown on one page and opposite is the solution to the question.

The first example shows a question in which the outline is partly completed to save time. The solution has thin lines for the parts which were given on the paper and thicker lines for those parts of the drawing which had to be completed by the candidate. This is a fairly straightforward question except for those parts of the assembly around the left hand end of the lever handle and the corresponding cut-out in the gear bracket. The question has to be studied closely to work out the arrangement of parts on the sectioned view in this region.

The second example is interesting because of the number of parts that have to be sectioned, each requiring a different section line. There are so many parts to be sectioned (nine) that inevitably one runs out of different types of section lines. Where the same shading is used twice, the lines have been used on parts that are some distance apart on the drawing. One could argue about whether or not to section the spindle; normally a spindle is not sectioned but this one is complicated enough in outline to perhaps justify sectioning. In fact it has been shown unsectioned.

The third example presents an assembly which, without the pictorial illustration, would be difficult to work out. Part 6 has some interesting cut-outs.

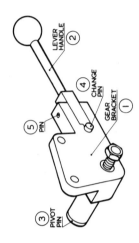

ALL DIMENSIONS ARE IN MILLIMETRES

LATHE GEAR CHANGE LEVER
THIRD ANGLE PROJECTION

⑤ PIN

② LEVER HANDLE

④ CHANGE PIN

③ PIVOT PIN

① GEAR BRACKET

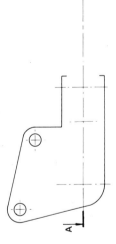

② LEVER HANDLE

④ CHANGE PIN

① GEAR BRACKET

⑤ PIN

③ PIVOT PIN

Example 1

The drawings (Fig. 18/36) show:
- (i) A pictorial sketch of a lathe gear change lever to act as a guide to show how its parts are fitted together.
- (ii) Details of the various parts of the gear change lever.
- (iii) Partly finished views of the gear bracket and lever handle of the lathe gear change lever.

You are required to do the following:
- (a) Complete the front view by adding the parts to make the whole assembly.
- (b) Complete the sectional plan on A–A.
- (c) Complete the end view.
- (d) Print the title GEAR CHANGE LEVER in letters 6 mm high.

Do not show any hidden detail.
  Dimension the following:
- (i) The diameter of the knob of the lever handle.
- (ii) The total length of the lever handle.
- (iii) The two 8 mm holes.

Associated Lancashire Schools Examining Board

Fig. 18/36 Specimen examination question

THIRD ANGLE PROJECTION

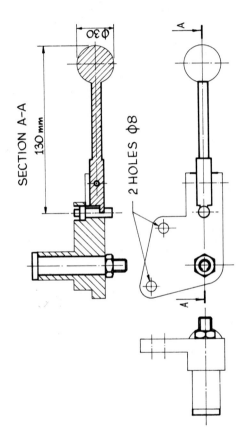

SECTION A–A

ϕ 30

130 mm

2 HOLES ϕ 8

A

A

GEAR CHANGE LEVER

Fig. 18/37   Scaled solution to Fig. 18/36

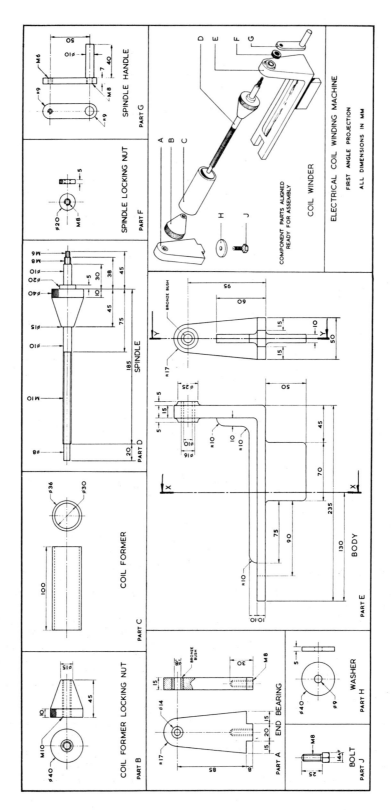

Fig. 18/38 Specimen examination question

226

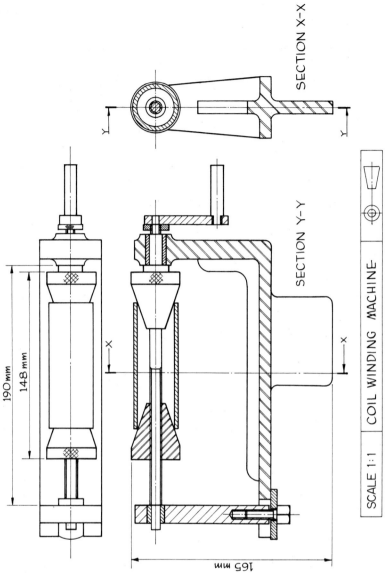

SECTION X-X

SECTION Y-Y

190 mm

148 mm

165 mm

| SCALE 1:1 | COIL WINDING MACHINE |

Fig. 18/39 Scaled solution to Fig. 18/38

Example 2
Electrical Coil Winding Machine

A pictorial view and details of each component part of a Coil Winding machine are shown (Fig. 18/38). A clip which fastens the wire to the Coil Former has not been included.

Draw, full size, either in First or Third Angle Projection the following views of the Unit completely assembled.

(i) A sectional front elevation taken on the cutting plane YY. Shown in Body detail Part E.
(ii) A sectional end elevation taken on the cutting plane XX. Shown in Body detail Part E.
(iii) A plan projected from view (i).

Hidden detail is not required in any view.

Use your own judgement to determine the size of any dimension not given.

A dimension shown as M10, for example, should be understood as

M means Metric Thread.

10 means Diameter of the Shaft or Hole in mm.

Make sure that the views are correctly positioned and in correct projection before drawing in any detail.

Credit will be given for good draughtsmanship and layout as well as for correct answers.

(i) Print in the title—Coil Winding Machine—size of letters to be 7 mm high.
(ii) Print in the scale and the system of projection used.
(iii) Put in the following dimensions:
   (a) the overall height of the Assembled Machine,
   (b) the length between the outside ends of the Coil Former Locking Nuts.
   (c) the length between the inner faces of the End Bearing Brackets.

Associated Lancashire Schools Examining Board

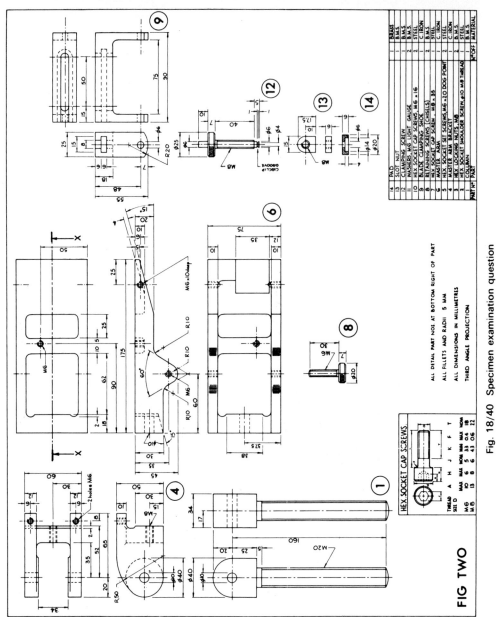

Fig. 18/40 Specimen examination question

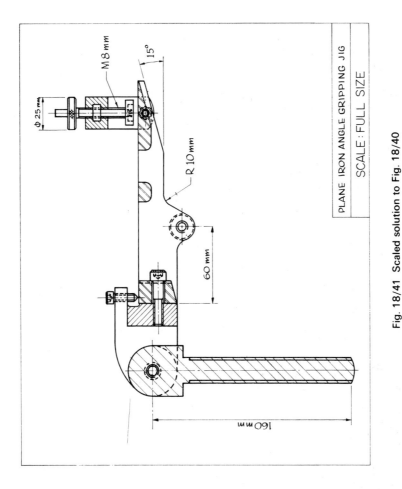

M 8 mm

15°

Φ 25 mm

R 10 mm

60 mm

160 mm

PLANE IRON ANGLE GRIPPING JIG

SCALE : FULL SIZE

Fig. 18/41 Scaled solution to Fig. 18/40

Example 3

Figure 1 (Fig. 18/40) is a pictorial view of a jig used to hold woodworker's plane irons and chisels at a set angle to the flat wheel surface of a grinding machine.

The jig is shown in position for grinding a plane iron.

For chisels, the jig is swivelled so that the master arm (6) is the other way up, and the blade clamping shoe (9) is swung round so that the chisel may be clamped.

The grinding angle may be altered by screwing the column (1) up or down.

The blade is moved to and fro across the face of the wheel by means of the operating lever which engages on the spigot on the clamping screw (12). For the purpose of this examination, the lever has been omitted.

Figure 2 shows detail drawings, **in third angle projection**, of the components of the plane iron grinding angle jig.

DRAW, FULL SIZE, A SECTIONAL ELEVATION ON THE SECTION PLANE X–X, with all the components assembled and the clamping screw (12) on the section plane and to the right of the view. The column (1) should be perpendicular, the master arm unit (4 & 6) horizontal and the blade clamping screw (12) perpendicular. In this view **do not** section the slot nut (13) and pad (14).

Correctly dimension: a horizontal measurement,
a vertical measurement,
an angular measurement (angle),
a radius,
a diameter, and
a screw thread.

Correctly position a suitable title block and state the following information:

Title: PLANE IRON GRINDING ANGLE JIG
Scale: FULL SIZE
Southern Universities' Joint Board

229

Exercises 18

(All questions originally set in Imperial units)

1. Fig. 1 shows a detail from a stationary engine. Draw this detail with the parts assembled. You may use either first-angle or third-angle projection.

Draw, *Twice full size:*

(a) A sectional front elevation in the direction of arrow B. The section should be parallel to the sides of the rod and pass through the centre of the hinge bolt;

(b) A plan in the direction of arrow A and projected from the elevation. Show all hidden detail in this view. (i) The smallest diameter spigot on the bolt should be shown threaded M10 for 15 mm and the bolt should be fastened by an M10 nut; (ii) Six main dimensions should be added to the views; (iii) Print the title 'Safety valve operating link detail' in the bottom right-hand corner of your sheet in 6 mm letters; (iv) In the bottom left-hand corner of the sheet print the type of projection you have used.

East Anglian Examinations Board

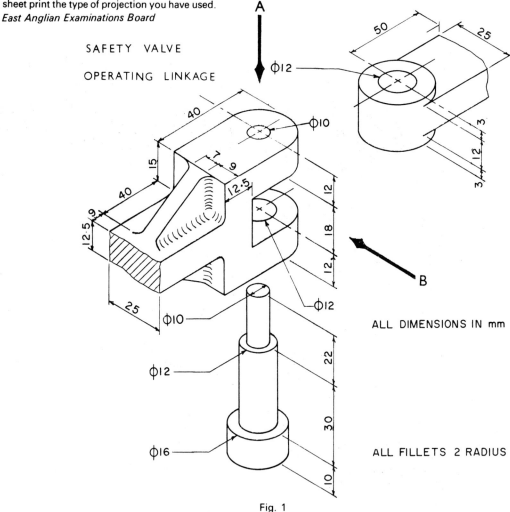

SAFETY VALVE

OPERATING LINKAGE

ALL DIMENSIONS IN mm

ALL FILLETS 2 RADIUS

Fig. 1

230

2. Fig. 2 shows in pictorial projection the proportions of a miniature camera with a lens hood attached. The lower drawing shows in detail the dimensions of the viewfinder, the rear face of which is vertical.

Using either first or third angle orthographic projection, draw full size the following views: (a) An elevation in the direction of arrow A; (b) A plan in the direction of arrow B.

Metropolitan Regional Examinations Board

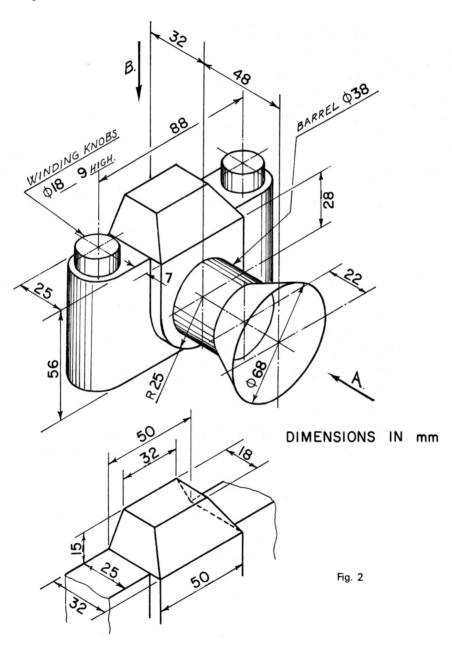

DIMENSIONS IN mm

Fig. 2

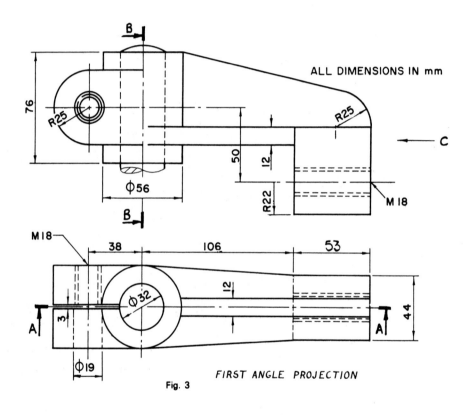

ALL DIMENSIONS IN mm

FIRST ANGLE PROJECTION

Fig. 3

3. Fig. 3 shows two views of a gear-change lever bracket from a lathe. When in use, the bracket is clamped to the 32 mm dia axle by a bolt passing through the 19 mm dia hole and screwing into the M18 threaded hole. The bolt is hexagon-headed, 62 mm long, with the thread running for 44 mm. With a portion of the axle in position and the bolt screwed tight, draw the following views of the bracket, full-size, in either first or third angle projection: (a) the given plan; (b) a sectional front view on the section line A–A; (c) a sectional end view on the section line B–B; (d) an end view in the direction of arrow C. Show hidden detail in view (a) only.

Insert, in a title block in the bottom right hand corner of your paper, the title GEAR-CHANGE LEVER BRACKET, the scale and projection used. Any dimensions not shown are left to your discretion.

Note: The bracket is a casting and the fillets and radiused corners have been left out to simplify the drawing. *Do not insert fillets and radiused corners in your drawing.*

Middlesex Regional Examining Board

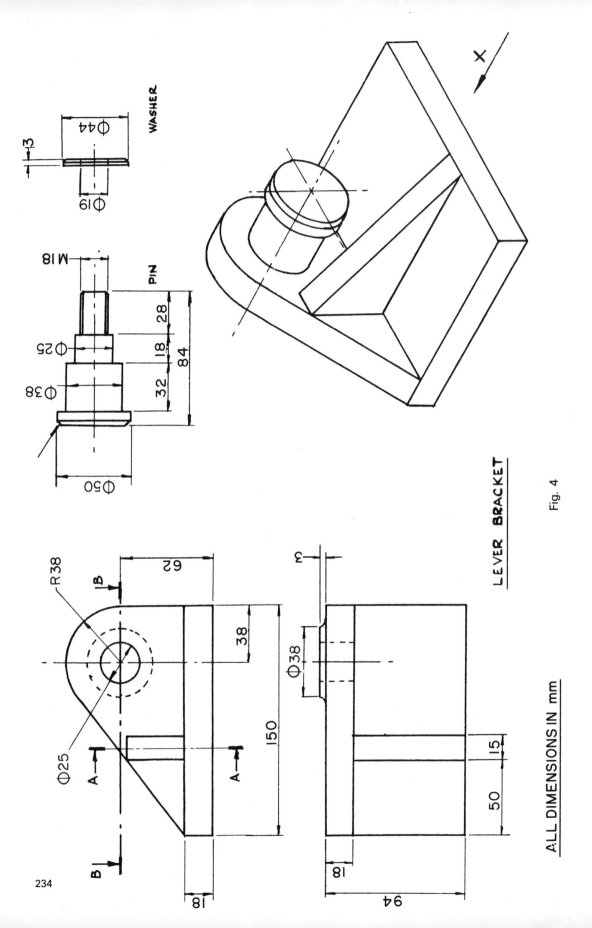

WASHER

PIN

LEVER BRACKET

Fig. 4

ALL DIMENSIONS IN mm

234

4. A sketch and views of a Lever Bracket are given in Fig. 4. The pin fits into the hole in the bracket and is held in position by means of an M18 Hexagon Nut.

Draw, full size, the following views of the assembled bracket, including the nut. (a) A front elevation looking in the direction of arrow X; (b) A sectional end view on AA looking in the direction of the arrows; (c) A sectional plan on BB looking in the direction of the arrows.

Hidden details need not be shown on the sectional views.

Insert the following four dimensions: (1) The length of the bracket; (2) The overall height of the bracket; (3) The overall width of the bracket; (4) The largest diameter of the pin.

Add a title and scale and state which method of projection you have used. Indicate also on your drawings the section planes AA and BB.

West Midlands Examinations Board

5. Fig. 5 shows an exploded view of a Tool Rest Holder of a wood turning lathe. The drawing shows the bolt, washer, the main casting and the clamp plate. Draw, full size, in first- or third-angle projection, an assembly drawing of all these parts as follows: (a) A sectional front elevation seen in the direction of arrow X taken along the centre line of the main casting; (b) A complete plan projected from view (a).

Hidden detail need not be shown. Insert the following dimensions: The width and thickness of the clamp plate; The width of the main casting; The distance between the centres of the slot; Print the title: 'TOOL REST HOLDER AND CLAMP'; State the system of projection used.

North Western Secondary School Examinations

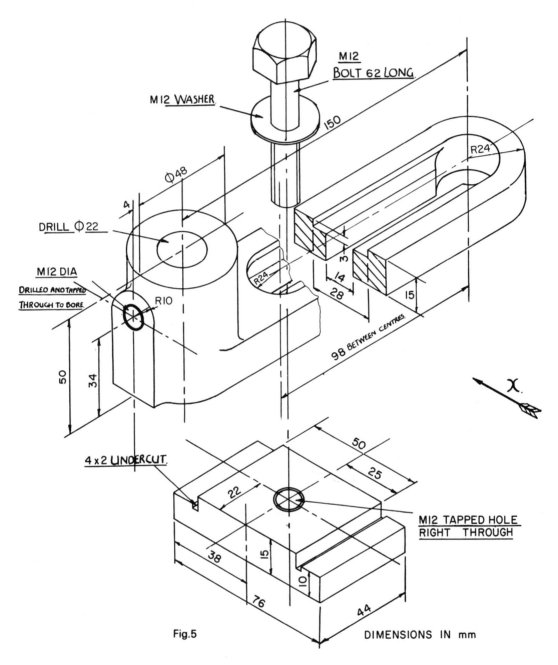

M12
BOLT 62 LONG.

M12 WASHER.

150

R24'

Φ48

DRILL Φ22

M12 DIA
DRILLED AND TAPPED
THROUGH TO BORE.

R24

R10

3

14

28

15

98 BETWEEN CENTRES.

50

34

X.

4 x 2 UNDERCUT.

50

25

22

M12 TAPPED HOLE
RIGHT   THROUGH

38

15

10

76

44

Fig.5

DIMENSIONS IN mm

236

6. Part A. With the items correctly assembled draw, *twice full size*, in either 1st or 3rd angle orthographic projection: (1) a sectional front view on the cutting plane A–A; (2) a plan. Do not show hidden detail of the nut.

Part B. Print, using block capital letters: (1) the title LEVER SUB-ASSEMBLY; (2) the scale used for your drawing; (3) the projection you have used.

Insert on your drawings *six main* dimensions in accordance with the standard method of dimensioning.

Assume any details or dimensions not given. Southern Regional Examinations Board

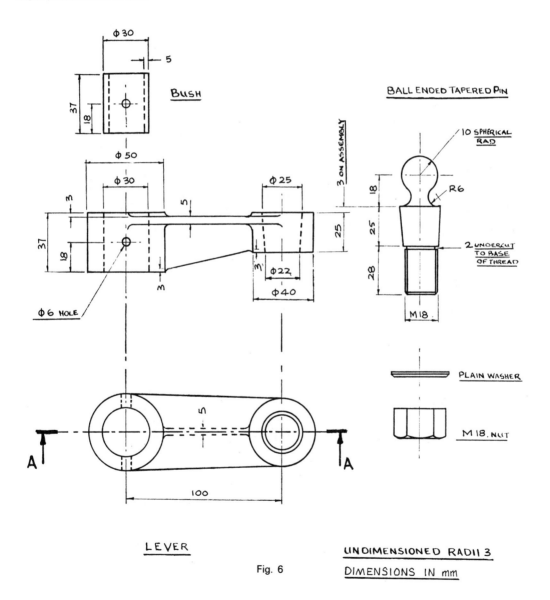

LEVER

Fig. 6

UNDIMENSIONED RADII 3

DIMENSIONS IN mm

237

7. Fig. 7 shows the details of the parts of a MACHINE
VICE. The movable jaw rests on the bed of the vice.
The thumbscrew is screwed into the body of the vice
and enters the 8 mm hole in the movable jaw.
The two pegs are fitted into the two holes in the movable jaw and secure in position the thumbscrew by
the 5 mm diameter neck.
(a) Draw *twice full size* the assembled vice with the
jaws 12 mm apart and the Tommy bar of the thumbscrew in the vertical position. (1) A front elevation
looking in the direction of arrow A; (2) A plan; (3)
An elevation looking in the direction of arrow B.
First or third angle projection may be used.
Hidden details need not be shown and only five
dimensions need be shown; (b) Make a good
quality *freehand* sketch of the assembled vice. The
sketch to be of approximately twice full size.
Add the TITLE, the SCALE and the ANGLE OF
PROJECTION used, in letters of a suitable size.
Southern Regional Examinations Board

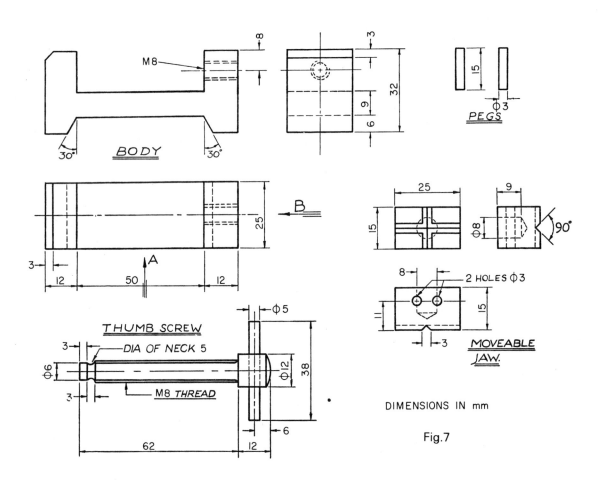

DIMENSIONS IN mm

Fig.7

238

8. Parts of a step-down pulley are shown in Fig. 8.
Draw, full size, the following views of the step-down
pulley fully assembled. (a) A sectional front elevation
on the plane XX; (b) An end elevation when viewed
in the direction of arrow A; (c) A plan projected from
(a) above and looking in the direction of arrow D.
Print the title. Size of letter to be 8 mm.
Print in the angle of projection you have used. Size of
letter to be 6 mm.
Print in the scale. Size of letter to be 4 mm.
No dimensions are required on your drawing. (1)
Radii at B = 12 mm. (2) Radii of Fillets = 6 mm; (3)
Use your own judgement to determine the size of any
dimensions not given on the drawing; (4) No hidden
details are to be shown.
Associated Lancashire Schools Examining Board

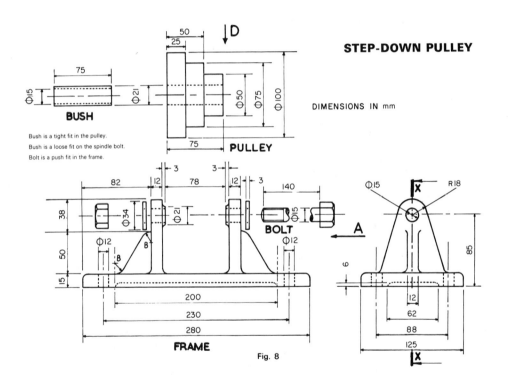

Fig. 8

9. Details of a cable roller are given in Fig. 9. Draw, *full-size*, the following views of the roller assembled in its stand: (a) a front elevation; (b) an end elevation in projection with the front view, and as seen from direction 'B'.

Show all hidden details.

Insert four important dimensions.
Beneath the views, print the title CABLE ROLLER in 5 mm capitals.
State which angle of projection has been used.
South-East Regional Examinations Board

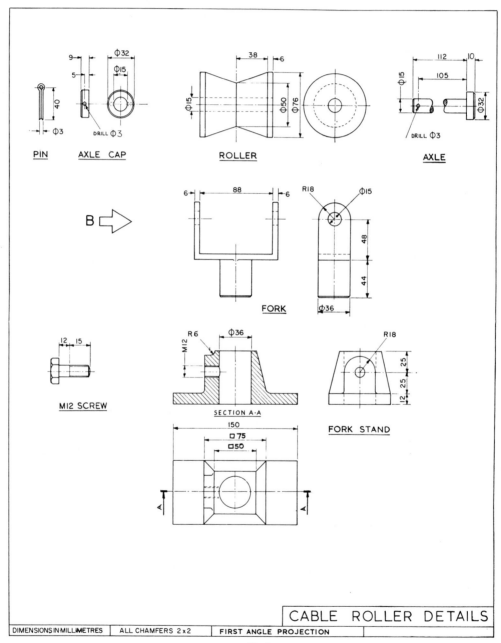

Fig. 9

10. Fig. 10 shows two views, in first angle projection, of the body and base of a SHAFT BEARING. The body is secured in position on the base by two hexagon-headed M8 set screws, 25 mm long, complete with washers. Draw, full size and in correct projection, the following views of the complete assembly, together with a length of 25 mm diameter shaft in position, approximately 150 mm long and shown broken at both ends: (a) A sectional side elevation on X–X in the direction indicated; (b) An elevation projected to the right of (a), the left-hand half to be sectioned on Y–Y; (c) A full plan projected from (a).

Draw in either first or third angle projection stating the method adopted.

Use your own discretion where any detail or dimension is not given.

Do not dimension your drawing; hidden edges are only required on view (c).

In a convenient corner of your paper draw a title block, 115 mm by 65 mm and print in it the drawing title, scale and your name.

University of London School Examinations

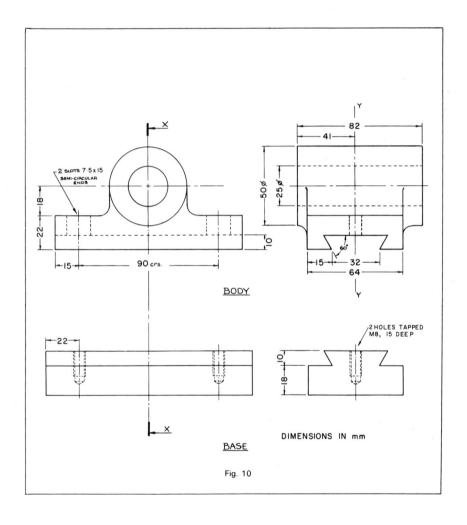

Fig. 10

11. Fig. 11 shows a light casting for a gear-lever bracket. Draw, full size : (a) A sectional elevation on X–X ; (b) An end elevation looking in the direction of arrow Y. Show *four* leading dimensions.

Hidden surfaces are not to be indicated. Suitable radii are to be assumed where not given.

Oxford and Cambridge Schools Examination Board

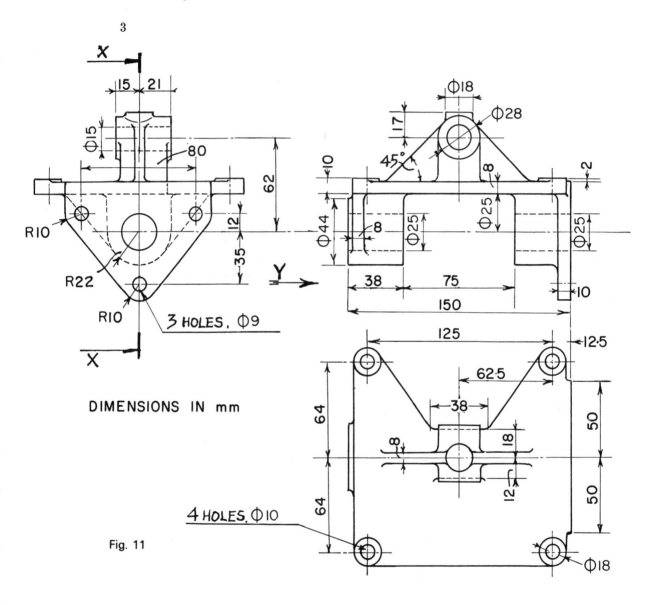

DIMENSIONS IN mm

Fig. 11

12. A special Relief Valve is shown in Fig. 12. The supply is at the valve 'D' and delivery is from the bore 'B'. Excess pressure is relieved through a valve at 'R'. Draw, full size, the following views: (a) An outside front elevation in the direction of the arrow 'F'; (b) A plan. The lower portion below the centre-line 'CC' being an outside view and the upper portion a section taken on the plane 'PP'; (c) A sectional end elevation on the plane 'EE'. Show hidden lines on view (c) only

Insert the following dimensions: (i) The distance between each valve and the delivery bore; (ii) The diameter of the flange; (iii) The distance between the face of the flange and the centre-line of the valves.

Complete in a suitable title block along the lower border of the paper, the title, scale, system of projection used and your name.

Oxford and Cambridge Schools Examination Board

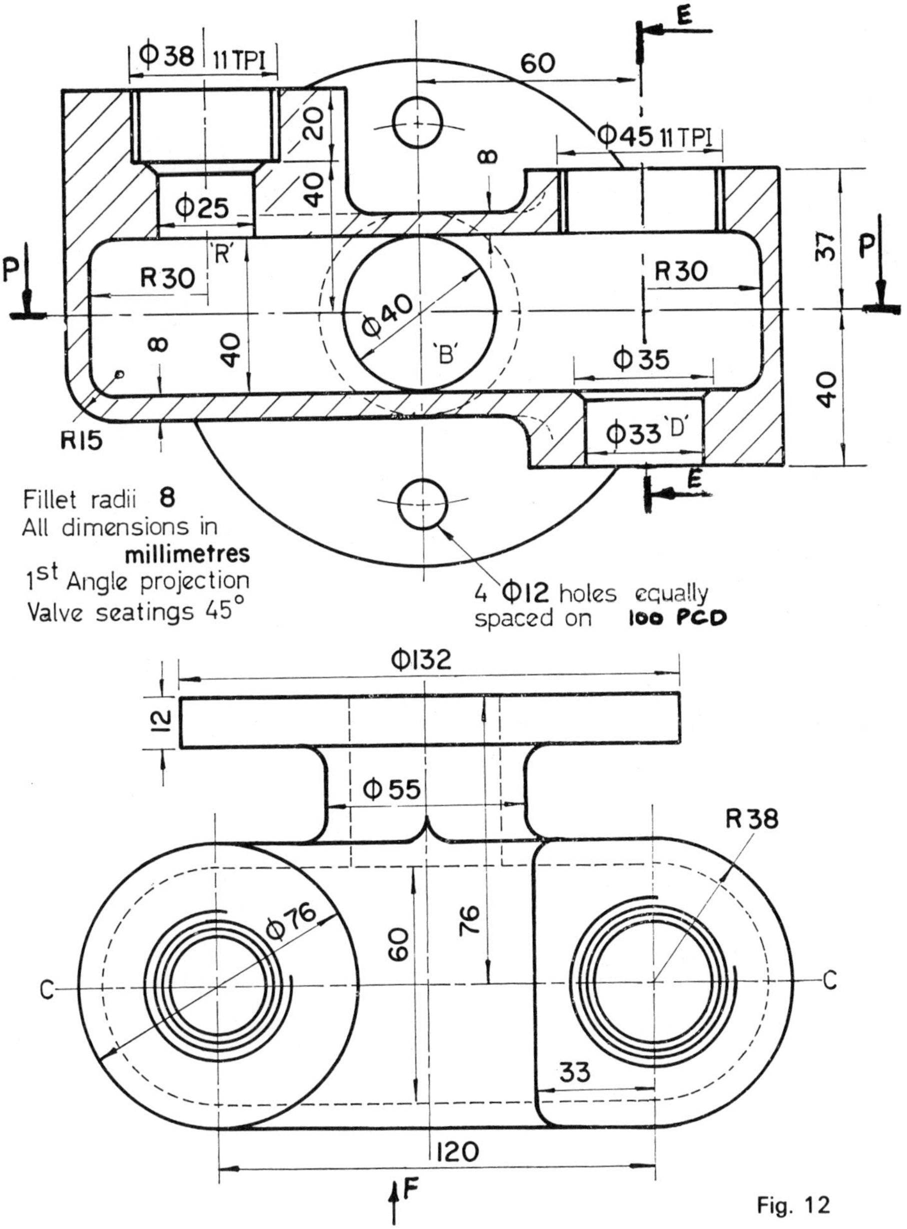

Φ38 11TPI
60
E
20
40
8
Φ45 11TPI
Φ25
'R'
P
R30
R30
37
P
Φ40
'B'
8
40
Φ35
40
Φ33 'D'
R15
E
Fillet radii 8
All dimensions in millimetres
1st Angle projection
Valve seatings 45°
4 Φ12 holes equally spaced on 100 PCD
Φ132
12
Φ55
R38
Φ76
60
76
C
C
33
120
F

Fig. 12

13. Orthographic views of a casting, drawn in third angle projection, are shown in Fig. 13. Do *not* copy the views as shown but draw full size in third angle projection the following: (a) a sectional elevation, the plane of the section and the direction of the required view being indicated at X–X; (b) an end elevation as seen in the direction of arrow E; (c) a complete plan as seen in the direction of arrow P and in projection with view (a).

Hidden edges are *not* to be shown on any of the views. Insert *two* important dimensions on each view, and in the lower right hand corner of the drawing paper draw a title block 115 mm by 65 mm and insert the relevant data.

Cambridge Local Examinations

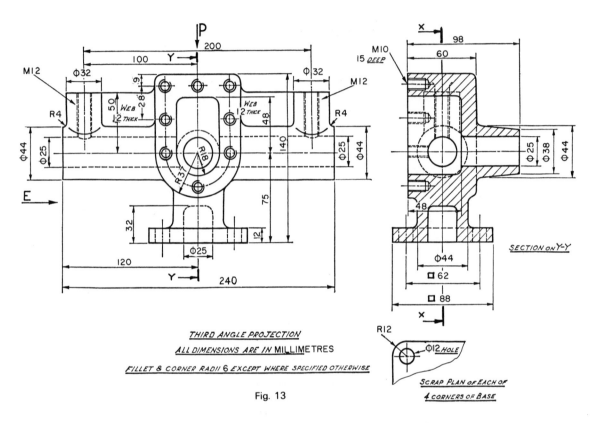

THIRD ANGLE PROJECTION
ALL DIMENSIONS ARE IN MILLIMETRES
FILLET & CORNER RADII 6 EXCEPT WHERE SPECIFIED OTHERWISE

Fig. 13

SECTION ON Y-Y

SCRAP PLAN OF EACH OF
4 CORNERS OF BASE

14. Fig. 14 shows parts of a magnifying glass as used by engravers and biologists. The stem A screws to the base B. The stand and the glass C are connected by two link bars D which are held in the desired position by the distance piece F and two 6 mm round headed bolts and wing nuts E.

Draw, full size, the following views of the magnifying glass and stand fully assembled: (a) an elevation in the same position as the encircled detail; (b) a plan projected from this elevation; (c) an elevation as seen in the direction of the arrow.

The link bars are inclined at 15° and the glass at 30° to the horizontal.

Include on this drawing eight of the main dimensions and a title block in the bottom right-hand corner. Within this block letter the title, MAGNIFYING GLASS AND STAND and state the scale and your school and name.

Southern Universities' Joint Board

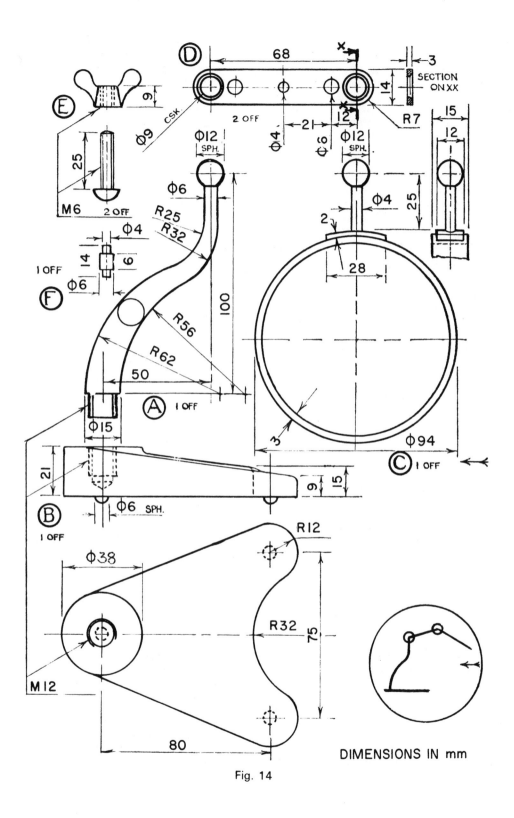

Fig. 14

DIMENSIONS IN mm

245

15. Fig. 15 is an exploded view of a plummer block bearing. Draw, to a scale of 2:1 in first angle orthographic projection, the following views of the assembled bearing: (a) a sectioned elevation as seen looking in the direction of arrow X; the cutting plane to be vertical and to pass through AA; (b) an elevation as seen when looking from the left of view (a); (c) a plan (beneath view (a)).

The cap is held in position on the studs by means of single chamfered hexagonal nuts that have 2 mm thick single bevelled plain washers beneath them.

Although the tapped hole, for the lubricator, is shown in the cap, further details of this hole are not shown and have been left to your own discretion. Details of this hole are required to be shown as are the details of the stud holes in the base.

Hidden detail of the base, *only*, is to be shown in view (c); no other hidden detail is to be shown.

Fully dimension the bottom brass, *only*.

Draw a suitable frame around your drawing and insert the title PLUMMER BLOCK, your name, the scale and the system of projection.

A parts schedule incorporating the part number, name of part and the number off of each part is to be completed in the lower right-hand corner of your sheet of drawing paper immediately above your title block.

Oxford Local Examinations

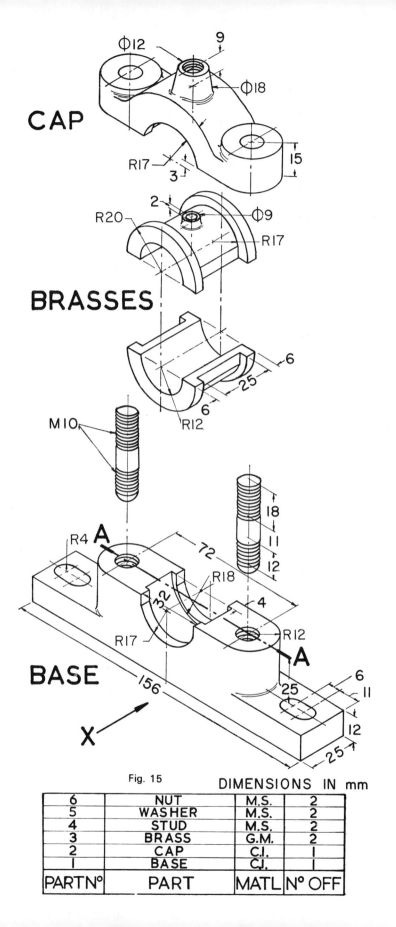

CAP

Φ12
9
Φ18
R17
3
15

BRASSES

2
R20
Φ9
R17
6
25
6
R12

M10

18
11
12

R4 A
72
32
R18
4
R17
R12
A

BASE
156
6
11
25
12
25

X

Fig. 15

DIMENSIONS IN mm

6	NUT	M.S.	2
5	WASHER	M.S.	2
4	STUD	M.S.	2
3	BRASS	G.M.	2
2	CAP	C.I.	1
1	BASE	C.I.	1
PART N°	PART	MATL	N° OFF

247

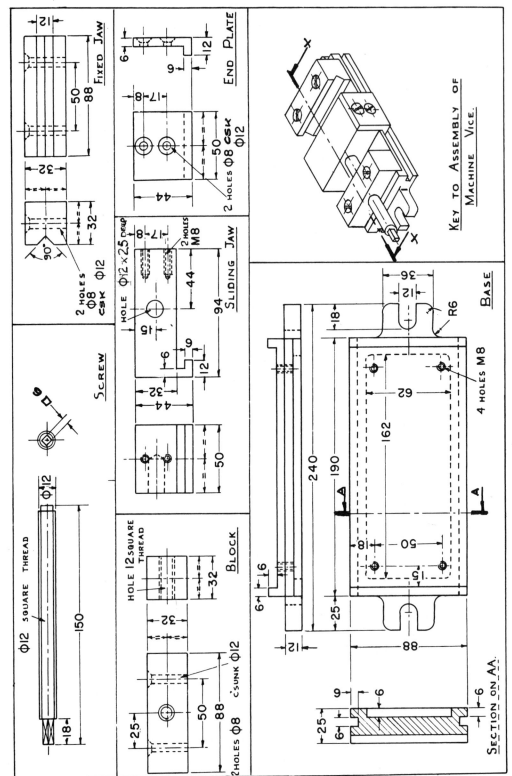

Fig. 16

248

16. Fig. 16 shows the details of a small machine vice and the key to its assembly. Draw full-size and in correct orthographic projection the following views of the completely assembled vice, the sliding jaw being approximately 25 mm from the fixed jaw. (a) A sectional elevation on a vertical plane passing through the axis of the square-headed screw, in the direction indicated by XX in the key; (b) A plan projected from the above.

Details of screws are not given, and these may be omitted from your drawing. No hidden edges are to be shown and dimensions are not required.

Either first-angle or third-angle (but not both) methods of projection may be used; the method chosen must be stated on the drawing.

In the bottom right-hand corner of the paper draw a title block 112 mm × 62 mm and in it print neatly the drawing title, MACHINE VICE ASSEMBLY, the scale and your name.

University of London School Examinations

17. Fig. 17A consists of a half-sectioned front elevation, a
    side elevation and a plan of part of a lathe steady.
    Draw, full size, in first angle orthographic projection,
    detail drawings of each of the six parts of the lathe
    steady, as follows:

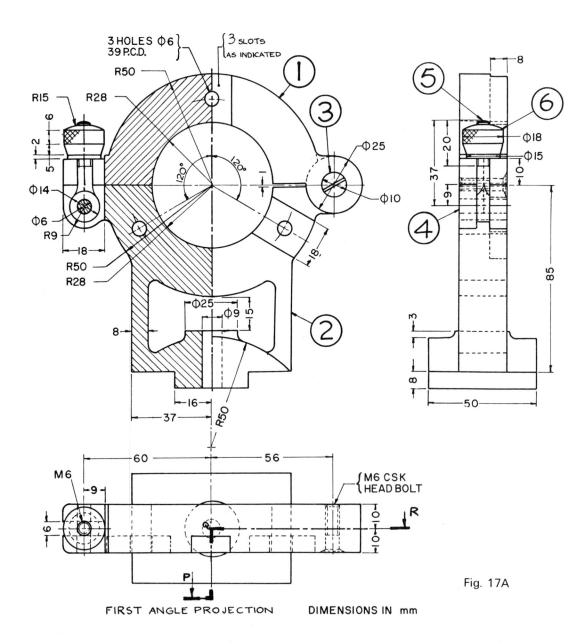

Fig. 17A

FIRST ANGLE PROJECTION   DIMENSIONS IN mm

Part 1. TOP. A front elevation and an elevation as seen from the right of the front elevation.

Part 2. BASE. A front elevation, a plan and a half-sectioned side elevation as seen from the right of the front elevation; the section planes for this view are PQ, QR.

Part 3. PIVOT SCREW. An elevation with the axis horizontal and the head to the right.

Part 4. HINGE PIN. An elevation with the axis horizontal and an end elevation as seen from the left.

Part 5. EYE BOLT. A front elevation and an elevation as seen from the left of the front elevation.

Part 6. CLAMPING NUT. A plan and elevation.

Full hidden detail is to be shown in Part 1 only; no other hidden detail is to be shown.

Part 5 is to be fully dimensioned; no other dimensions are to be shown.

Draw a suitable frame around your drawing and insert, in accordance with the recommendations of B.S. 308, the title LATHE STEADY, your name, the scale and the system of projection.

Also insert, in the right-hand corner of the frame, above the title block, a parts list giving the part no., name of part and the No. off. It is suggested that the parts list should have a width of 125 mm.

The lower part of Fig. 17B gives an indication of how your sheet should be arranged.

Oxford Local Examinations

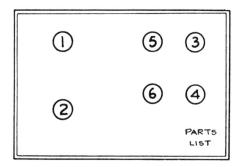

Fig. 17B

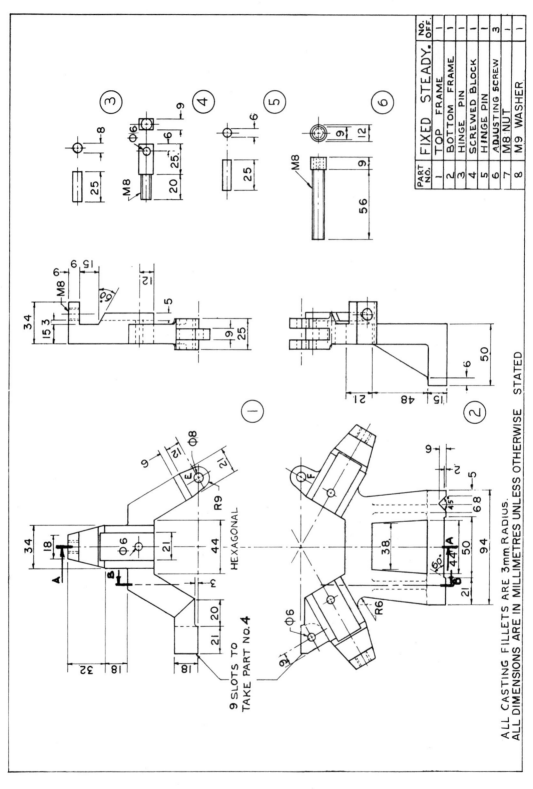

PART NO.	FIXED STEADY.	NO. OFF.
1	TOP FRAME	1
2	BOTTOM FRAME	1
3	HINGE PIN	1
4	SCREWED BLOCK	1
5	HINGE PIN	1
6	ADJUSTING SCREW	3
7	M8 NUT	1
8	M9 WASHER	1

ALL CASTING FILLETS ARE 3mm RADIUS.
ALL DIMENSIONS ARE IN MILLIMETRES UNLESS OTHERWISE STATED

Fig. 18

9 SLOTS TO TAKE PART NO. 4

HEXAGONAL

252

18. The detail drawings in Fig. 18 are parts of a fixed steady. The adjustable bearing blocks, screws to hold them in position and the means of fixing the steady to the lathe have been omitted.

When the steady is assembled, the top frame hinges on the right-hand lug and turns about part number 3. The left-hand slot of the bottom frame takes part number 4 which turns about part number 5 into the left-hand slot of the top frame. The two frames, parts 1 and 2, are held together by an M8 nut which is screwed with a washer on part 4.

To a scale of three-quarters full size draw the follow-

ing views of the assembled steady: (a) an elevation looking perpendicularly to the hexagon, i.e. in a similar position to the given left-hand views; (b) project a plan from the elevation; (c) a sectional elevation on AA; (d) a sectional elevation on BB.

Only show one adjusting screw, part number 6, this to be fully screwed into the lug below A. Hidden detail is not required for the sectional elevations. Add six main dimensions. In block letters give the following: FIXED STEADY; SCALE; and NAME.

Southern Universities Joint Board

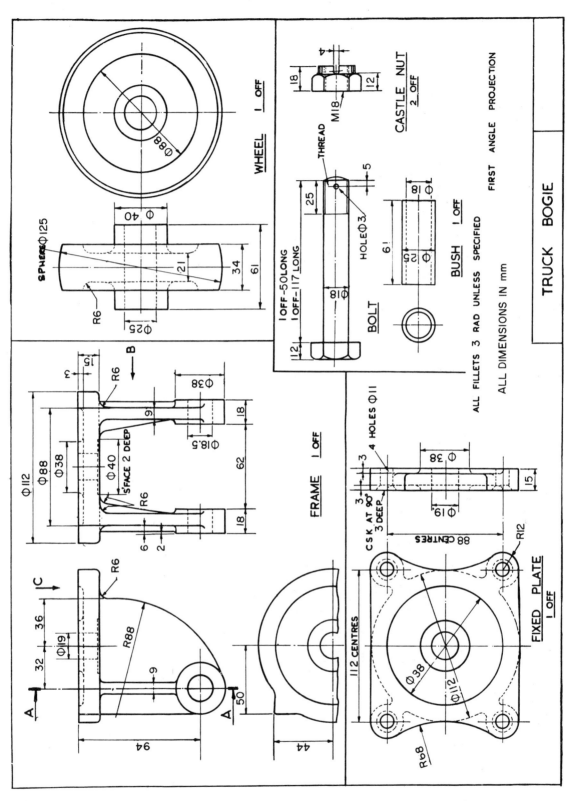

WHEEL 1 OFF

CASTLE NUT
2 OFF

M18

THREAD

HOLE Φ3

BOLT

1 OFF—50LONG
1 OFF—117 LONG

BUSH 1 OFF

FIRST ANGLE PROJECTION

ALL FILLETS 3 RAD UNLESS SPECIFIED

ALL DIMENSIONS IN mm

TRUCK BOGIE

SPHERE Φ125

R6

FRAME 1 OFF

S'FACE 2 DEEP

R6

4 HOLES Φ11

C S K AT 90°
3 DEEP

88 CENTRES

112 CENTRES

R88

R6

FIXED PLATE
1 OFF

Fig. 19

254

19. Fig. 19 opposite shows details of a bogie truck. The fixed plate is secured to the truck by four M10 countersunk screws (not to be shown) and carries the wheel and frame by means of a central bolt M18, 50 mm long. With the parts correctly assembled and allowing a 2 mm clearance between the top of the frame and the underside of the plate draw, *full-size*, the following views: (a) a sectional front elevation looking in the direction of arrows AA and taken on the centre-line as indicated; (b) an outside end elevation looking in the direction of arrow B, the longer side of the fixed plate to be shown in this view; (c) an outside plan projected from (b) above and looking in the direction of arrow C.

Hidden details to be shown in view (b) *only*.

The castle nuts to be shown in view (b) *only*.

Insert the following dimensions: (i) the distance between the centres of the fixing screw holes in the fixed plate in both directions; (ii) the distance of the centre of the wheel from the top of the frame; (iii) the outside diameter of the wheel; (iv) the internal diameter of the bush.

Add, in letters 10 mm high, the title TRUCK BOGIE, and in letters 6 mm high the scale and system of projection used.

First-angle or third-angle projection may be used but the three views must be in a consistent system of projection.

Associated Examining Board

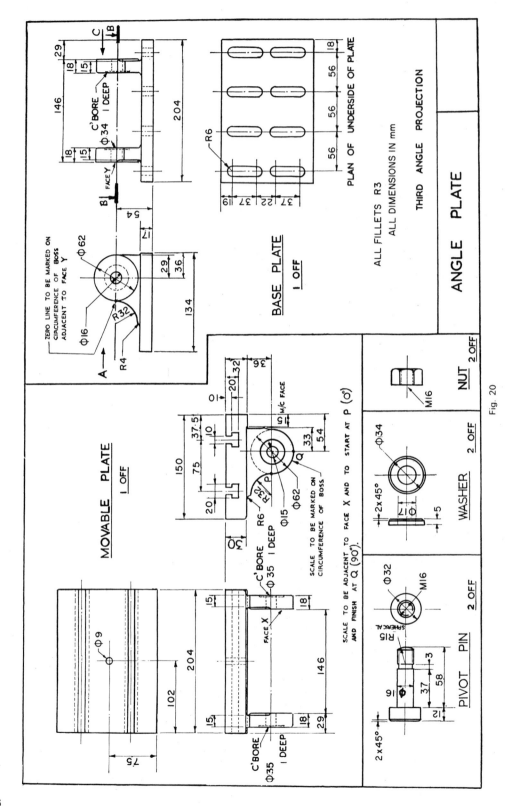

BASE PLATE 1 OFF

PLAN OF UNDERSIDE OF PLATE

ALL FILLETS R3
ALL DIMENSIONS IN mm

THIRD ANGLE PROJECTION

ANGLE PLATE

ZERO LINE TO BE MARKED ON CIRCUMFERENCE OF BOSS ADJACENT TO FACE Y

MOVABLE PLATE 1 OFF

SCALE TO BE MARKED ON CIRCUMFERENCE OF BOSS

SCALE TO BE ADJACENT TO FACE X AND TO START AT P (0°) AND FINISH AT Q (90°).

NUT 2 OFF
M16

WASHER 2 OFF

PIVOT PIN 2 OFF
SPHERICAL
M16

Fig. 20

20. Fig. 20 opposite shows details in third-angle projection of an adjustable Angle Plate capable of moving through angles 0 to 90 degrees.

With the parts correctly assembled and arranged for an angle of 0 degrees, i.e. base plate and movable plate both horizontal, draw, full size, the following views: (a) a front elevation looking in the direction of arrow A.; (b) an end elevation looking in the direction of arrow C; (c) a sectional plan projected from (a), the section being taken on the centre-line through the pivot pins and looking in the direction of arrows BB. The pivot pins are driving fits in the base plate bosses and sliding fits in the movable plate bosses. The pins are to be assembled with their 32 mm diameter heads placed on the 34 mm counterbore sides of the bosses.

Hidden detail is to be shown in view (a) only.

Only the scale markings for 0, 30, 60 and 90 degrees in view (a) are to be shown.

Insert the following dimensions: (i) the length of the base plate; (ii) the width of the base plate; (iii) the vertical height of the centre of the pivot above the bottom face of the base plate; (iv) the diameter of the pivot pin; (v) the distance between the tee slots in the movable plate.

In a rectangle 150 mm long and 75 mm wide in a corner of the drawing insert, in letters 10 mm high, the title ANGLE PLATE and, in letters 6 mm high, the scale and system of projection used.

First-angle or third-angle projection may be used but the three views must be in a consistent system of projection.

Associated Examining Board

257

APPENDIX A

SIZES OF ISO METRIC PRECISION HEXAGON NUTS, BOLTS AND WASHERS

All dimensions in millimetres (maximum)

Size	Diameter of shank	Width across flats	Height of bolt head	Radius under head	Depth of washer face Washer thickness	Thickness of normal nut	Thickness of thin nut	Washer inside diameter	Washer outside diameter
		A/F	T	R	D	T_N	T_{TN}	d_1	d_2
M1.6	1 6	3.2	1 125	0.2	—	1 3	—	2 0	
M2	2.0	4.0	1.525	0.3	—	1.6	—	2.6	—
M2.5	2.5	5.0	1.825	0.3	—	2.0	—	3.1	—
M3	3.0	5.5	2.125	0.3	0.1	2.4	—	3.6	5.08
M4	4.0	7.0	2.925	0.35	0.1	3.2	—	4.7	6.55
M5	5.0	8.0	3.650	0.35	0.2	4.0	—	5.7	7.55
M6	6.0	10.0	4.15	0.4	0.3	5.0	—	6.8	9.48
M8	8.0	13.0	5.65	0.6	0.4	6.5	5.0	9.2	12.43
M10	10.0	17.0	7.18	0.6	0.4	8.0	6.0	11.2	16.43
M12	12.0	19.0	8.18	1.1	0.4	10.0	7.0	14.2	18.37
M14	14.0	22.0	9.18	1.1	0.4	11.0	8.0	16.2	21.37
M16	16.0	24.0	10.18	1.1	0.4	13.0	8.0	18.2	23.27
M18	18.0	27.0	12.215	1.1	0.4	15.0	9.0	20.2	26.27
M20	20.0	30.0	13.215	1.2	0.4	16.0	9.0	22.4	29.27
M22	22.0	32.0	14.215	1.2	0.4	18.0	10.0	24.4	31.21
M24	24.0	36.0	15.215	1.2	0.5	19.0	10.0	26.4	34.98
M27	27.0	41.0	17.215	1.7	0.5	22.0	12.0	30.4	39.98
M30	30.0	46.0	19.26	1.7	0.5	24.0	12.0	33.4	44.98
M33	33.0	50.0	21.26	1.7	0.5	26.0	14.0	36.4	48.98
M36	36.0	55.0	23.26	1.7	0.5	29.0	14.0	39.4	53.86
M39	39.0	60.0	25.26	1.7	0.6	31.0	16.0	42.4	58.86
M42	42.0	65.0	26.26	1.8	0.6	34.0	16.0	45.6	63.76
M45	45.0	70.0	28.26	1.8	0.6	36.0	18.0	48.6	68.76
M48	48.0	75.0	30.26	2.3	0.6	38.0	18.0	52.6	73.76
M52	52.0	80.0	33.31	2.3	—	42.0	20.0	56.6	—
M56	56.0	85.0	35.31	3.5	—	45.0	—	63.0	—
M60	60.0	90.0	38.31	3.5	—	48.0	—	67.0	—
M44	64.0	95.0	40.31	3.5	—	51.0	—	71.0	—
M68	68.0	100.0	43.31	3.5	—	54.0	—	75.0	—

ACKNOWLEDGMENT
(The values above are extracted from BS 3692: 1967, *Specification for ISO Metric Precision Hexagon Bolts, Screws and Nuts (Metric Units)*, and are reproduced by permission of the British Standards Institution, 2 Park Street, London, W.1, from whom copies of the complete standards may be obtained).

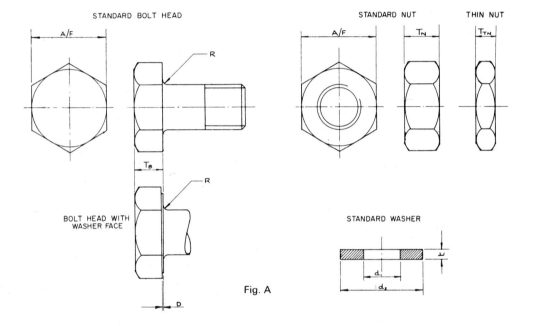

STANDARD BOLT HEAD

A/F

R

T_B

BOLT HEAD WITH
WASHER FACE

R

D

STANDARD NUT

A/F

T_N

THIN NUT

T_TN

STANDARD WASHER

t

d_1

d_2

Fig. A

259

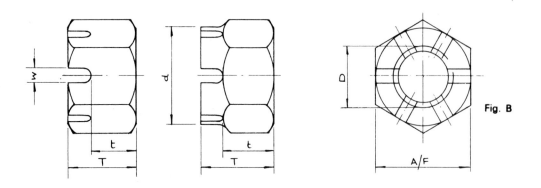

Fig. B

APPENDIX B

SIZES OF SLOTTED AND CASTLE NUTS WITH METRIC THREAD

All dimensions in millimetres

Thread diameter	Diameter of castellated portion	Total thickness	Thickness of hexagon portion	Width of slot	Width across flats
D	d	T	t	W	A/F
4	—	5.0	3.2	1.4	7
5	—	6.0	4.0	1.6	8
6	—	7.5	5.0	2.2	10
8	—	9.5	6.5	2.7	13
10	—	12.0	8.0	3.0	17
12	17	15.0	10.0	3.8	19
14	19	16.0	11.0	3.8	22
16	22	19.0	13.0	4.8	24
18	25	21.0	15.0	4.8	27
20	28	22.4	16.0	4.8	30
22	30	26.0	18.0	5.8	32
24	34	27.0	19.0	5.8	36
27	38	30.0	22.0	5.8	41
30	42	33.0	24.0	7.3	46
33	46	35.0	26.0	7.3	50
36	50	38.0	29.0	7.3	55
39	55	40.0	31.0	7.3	60
Thread diameter	Diameter of castellated portion	Total thickness	Thickness of hexagon portion	Width of slot	Width across flats

Sizes over φ 39 have eight slots

Dimensions above are for drawings only

Index